Electronics
is Easy?

Electronics is Easy?

What does industry need?

Clyde T. Eisenbeis

Electronic Design Engineer

Fox Tck, LLC

ISBN-13: 978-0-9963514-0-9
LCCN: 2015913206

Cover Design by Alan Pranke
Typeset by MK Ross

Printed in the United States of America

Preface

What does industry need?

Industry needs Electrical Engineers (EEs) who understand basic electronic fundamentals, are quick learners (technology changes), and have good people skills.

"Electronics is Easy?" covers basic fundamentals for Electronic Design, Test, Manufacturing, Application, Support, Sales, Marketing, and System Engineers. The contents strive to be clear and concise. The intent is to prepare EE students for the world of electronics.

This book covers details essential for electronics in industry. A solid understanding of electronic fundamentals is crucial for all electronic products.

SPICE models are an exceptional resource to assist with electronic design. A clear understanding of electronic fundamentals makes them easy to use. These models provide insight into the trade-offs for different components, and a solid understanding of how to improve the design.

A wide range of topics, ranging from design and documentation to test and manufacturing, are covered by the book. Understanding the contents of this book increases the odds of electronic products that perform as required.

This book covers the foundation. There is more to know to be truly proficient.

I hope you enjoy the book.

SPICE
Information about free SPICE programs and SPICE models can be found at http://ElectronicsIsEasy.com.

Acknowledgements

The people who helped me learn electronics is a long list. It started at the University of North Dakota and my first job at Texas Instruments. This continued when I worked at 3M, Turtle Mountain Corporation, and Fisher / Emerson Process.

Initially I acquired my knowledge from college faculty, coworkers, books, and magazines. Later, my knowledge was expanded by the Internet and SPICE models.

Early in my career, I did not realize that technology would change continuously. Electronic components change with higher performance, lower power, smaller size, and lower cost. It is challenging to stay current.

People who helped review this book include my wife Sheryl, and our sons, Kirk, Jeff, and Ross, who have EE degrees.

I am deeply indebted to my friends Bruce Trump (Texas Instruments), Mark Byer (Fisher / Emerson Process), Dennis Eisenbeis (Intuitive Surgical), and Bill Kimmel (Kimmel Gerke Associates) for their extensive reviews.

I also acknowledge the contributions of my friends Brett Bonin (Maxim), Phillip Lorenz (Lorenz Sales), Alex Toy (Medtronic), Xiaofan Yang (Qualcomm), Bob Rynkiewicz (Linear Technology), Mark Hopkins (Allegro MicroSystems), Andy Thompson (Spectrum Software)

Special thanks to Hossein Salehfar (University of North Dakota), Scott C. Smith (North Dakota State University), and Gary Tuttle (Iowa State University).

Mark Levine (Hillcrest Media Group, Inc.) has provided assistance with self-publishing. His book, "The Fine Print of Self-Publishing", has been a good resource.

Figures and plots were created using MicroCap (Spectrum Software).

Introduction

Electronics is often perceived as being complex and difficult to understand.

Many years ago I realized there is nothing difficult to learn. The hardest part is finding the right person to explain it.

This book attempts to document what I have learned as an Electronic Design Engineer.

Contents

Chapter 1

ELECTRONICS BASICS

1.1 Basics

The basics of electronics can be described with a garden hose.

The higher the water pressure, the more water flows through the hose. Closing the water faucet reduces the water pressure which causes less water to flow through the hose.

Reducing the size of the hose from 5/8 inch to 1/2 inch creates more resistance that causes less water to flow through the hose.

Electronics is the same. Voltage is like water pressure, current is like water, and a resistor is like the hose diameter.

Voltage causes current to flow through a resistor. A larger voltage causes more current to flow through the resistor. A larger resistance causes less current to flow through the resistor.

Ohm's Law defines the relationship between voltage (V), current (I), and resistance (R) (Figure 1.1-1).

Ohm's Law: $V=I*R$ ("*" indicates multiply)
$V=6V, R=2\Omega$
$I=6V/2\Omega=3A$

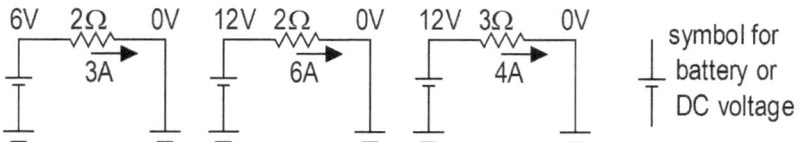

Figure 1.1-1

The standard unit for voltage is volts (V), for current is amps (A), and for a resistor is ohms (Ω). Common resistor value terminology is ohms, k ohms, and M ohms:

1   1Ω
1k   1,000Ω
1M   1,000,000Ω

Common current terminology is Amps, milliamps, microamps, and nanoamps:

1A   1 Amp
1ma (milliamp) 0.001 (10^{-3}) Amps
1µA (microamp) 0.000001 (10^{-6}) Amps
1nA (nanoamp)   0.000000001 (10^{-9}) Amps

1.2 Voltage (DC) and Ground

A common voltage for alkaline batteries is 1.5V. The nomenclature is to define the "-" on the battery as ground (Gnd) and the "+"
as voltage (V) (Figure 1.2-1).

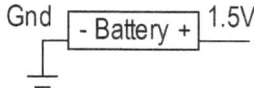

Figure 1.2-1

Power supplies also generate voltages. Some are fixed voltages,
such as 3V, 5V, and 12V. Some are variable (Figure 1.2-2).

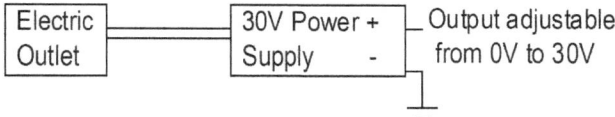

Figure 1.2-2

The output of a variable power supply can be adjusted to provide
a range of voltages (such as all voltages between 0 and 30V).

Power supplies and batteries generate a DC (direct current)
voltage. An ideal DC voltage power supply does not change.
Once set to 5V, the voltage output of an ideal power supply is
always 5V. Actual power supplies often have slight variations in
voltage output.

A perfect battery also provides a constant voltage. However,
over time, batteries deplete and the voltage begins to drop.

1.3 Voltage (AC)

While power supplies create a DC voltage, function generators
can generate an AC (alternating current) voltage.

An AC voltage can be described as a vibrating string on a
guitar. Pluck the string and the string vibrates. The vibration rate
is given as cps (cycles per second) or Hz (Hertz). It represents the
number of times the string vibrates in one second (Figure 1.3-1).

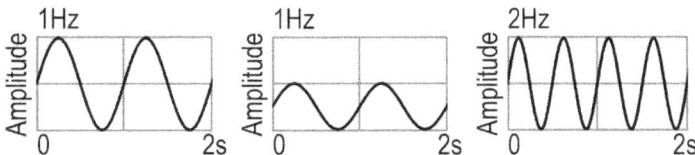

Figure 1.3-1

The word amplitude could be replaced by the word distance. Pluck the string softer and the string travels a shorter distance.

Shortening a guitar string causes the string to vibrate faster, known as a higher pitch. The numbers associated with pitch are known as frequency. If the string vibrates one time per second, it is known as 1Hz. If it vibrates two times per second, it is known as 2Hz.

Function generators can change the output amplitude and the output frequency.

An AC voltage across a resistor creates an AC current. The wave shape is the same. An AC voltage that oscillates from -10V to +10V across a 2Ω resistor results in current that oscillates from -5A to +5A (Figure 1.3-2).

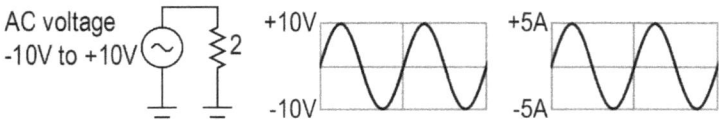

Figure 1.3-2

An AC voltage that oscillates from 0V to +10V across a 2Ω resistor results in current that oscillates from 0A to +5A (Figure 1.3-3).

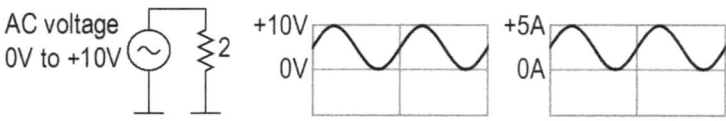

Figure 1.3-3

1.4 Electro-Magnetism

One of the most significant discoveries, ever, is electro-magnetism. Moving a magnet past a wire produces an electric current. The faster the motion, the larger the electric current.

This led the world to electricity. All coal-fired power plants, nuclear power plants, and hydro-electric power plants spin magnets past wires which results in electricity.

Likewise, current through a wire creates a magnetic field which causes a magnet to move, or the wire to move relative to a magnet. This led to motors.

Chapter 1 Quiz

R	resistor	V	voltage
C	capacitor	I	current
L	inductor	**T**	time constant

1-1   With 15V across a 10Ω resistor, what is the value of the current?

1-2   With 10V across a resistor and 5A through the resistor, what is the resistor value?

1-3   What is standard voltage for most alkaline batteries?

1-4   Do properly designed power supplies generate AC voltages?

1-5   What units represent the AC frequency?

1-6   What equipment is used to generate an AC voltage?

Chapter 2

Analog Passive Components

2.1 Electronic Component Physical Packages

There are two physical packages of electronic components; PTH (pin through hole) and SMT (surface mount technology). PTH and SMT are commonly used for resistors, capacitors, diodes, and inductors.

PTH components are attached to a PCB (printed circuit board) by bending the leads and inserting into PCB metal holes. Components are attached to the PCB using solder (Figure 2.1-1).

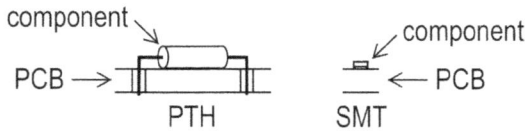

Figure 2.1-1

SMT components are attached by soldering to metal foils that are on the PCB surface. SMT components are smaller package sizes and do not interfere with components mounted on the opposite side of the PCB.

New technology, such as 3D printing, may change PCB fabrication. This technology is in the early development phase and may change a fabrication that has existed for a few decades.

2.2 Resistor Parameters

A resistor affects the flow of current. The larger the resistor value, the more it restricts the current.

Resistor parameters include value (ohms), tolerance (%), power rating (watts), and temperature coefficient (ppm/°C). All of these parameters may be important in a design.

Standard resistor tolerances are ±5%, ±1%, and ±0.1%. The standard values differ for each tolerance.

Values for 5% resistors (known as the E24 standard) are decade multiples of:

10 11 12 13 15 16 18 20 22 24 27 30
33 36 39 43 47 51 56 62 68 75 82 91

Examples of 5% decade multiples are 13Ω, 130Ω, 1.3kΩ, 13.kΩ, 130kΩ, and 1.3MΩ.

The most common resistors are ±1% (designated as 1%). Values for 1% resistors (known as the E192 standard) are decade multiples of:

10.0 10.2 10.5 10.7 11.0 11.3 11.5 11.8 12.1 12.4 12.7 13.0

13.3 13.7 14.0 14.3 14.7 15.0 15.4 15.8 16.2 16.5 16.9 17.4

17.8 18.2 18.7 19.1 19.6 20.0 20.5 21.0 21.5 22.1 22.6 23.2

23.7 24.3 24.9 25.5 26.1 26.7 27.4 28.0 28.7 29.4 30.1 30.9

31.6 32.4 33.2 34.0 34.8 35.7 36.5 37.4 38.3 39.2 40.2 41.2

42.2 43.2 44.2 45.3 46.4 47.5 48.7 49.9 51.1 52.3 53.6 54.9

56.2 57.6 59.0 60.4 61.9 63.4 64.9 66.5 68.1 69.8 71.5 73.2

75.0 76.8 78.7 80.6 82.5 84.5 86.6 88.7 90.9 93.1 95.3 97.6

Examples of 1% decade multiples are 1.05Ω, 10.5Ω, 105Ω, 1.05kΩ, 10.5kΩ, 105kΩ, and 1.05MΩ.

Resistor power ratings are important to reliability and performance of a product. A resistor can change value, or be destroyed, if it becomes too hot.

$V=4V, R=8\Omega, I=0.5A$
$P=V*I$
$P=4V*0.5A=2w \ (watts)$
$P=V^2/R$
$P=4V^2/8\Omega=16/8w=2w$
$P=I^2*R$
$P=0.5A^2*8\Omega=0.25*8w=2w$

The SMT physical size specifies the amount of power a resistor can handle. The type number indicates the approximate size.

type	size	max power
0402	0.040 x 0.020 inches	1/16 watt
0603	0.060 x 0.030 inches	1/10 watt
0805	0.080 x 0.050 inches	1/8 watt
1206	0.120 x 0.060 inches	1/4 watt
1210	0.120 x 0.100 inches	1/3 watt
2512	0.250 x 0.120 inches	1 watt

Resistor values change over temperature. The changes are specified as a temperature coefficient in ppm/°C.

1 ppm	1/1,000,000
1 ppm	1/10,000%
1 ppm	0.0001%
100 ppm	0.01%

Different materials have different temperature stability values. Temperature coefficients for thick film resistors range from ±200 to ±250 ppm/°C. Thin film resistors range from ±5 to ±25 ppm/°C.

The composition of resistors is some form of carbon and / or metal film. It is common for power resistors to be wound with nichrome wire which gives them a higher power rating.

2.3 Parallel / Series Resistors

A combination of parallel and series resistors are used when the required resistor value is not standard. Series and parallel resistor combinations are easy to calculate (Figures 2.3-1 and 2.3-2).

$$R_{series}=R1+R2$$
$$R_{series}=2k+4.99k=6.99k$$

Figure 2.3-1

$$R_{parallel}=(R1*R2)/(R1+R2)$$
$$R_{parallel}=(10k*10k)/(10k+10k)=5k$$
$$R_{parallel}=(49.9k*20k)/(49.9k+20k)=14.28k$$

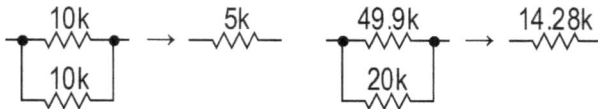

Figure 2.3-2

Resistors can work as voltage dividers (Figure 2.3-3).

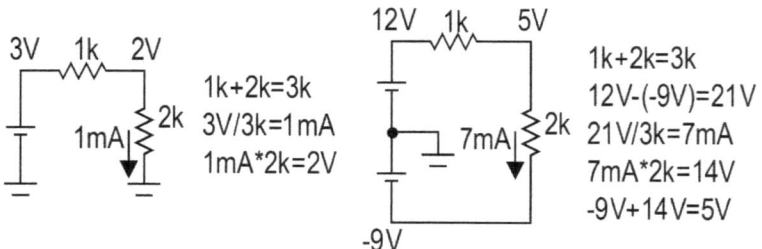

1k+2k=3k
3V/3k=1mA
1mA*2k=2V

1k+2k=3k
12V-(-9V)=21V
21V/3k=7mA
7mA*2k=14V
-9V+14V=5V

Figure 2.3-3

It is good practice to include tolerance percentages in calculations. For higher precision, 0.1% resistors can be used.

2.4 Capacitors (DC)

A capacitor is similar to a water bucket. Water fills a bucket;
current fills a capacitor with charge. Current is charge over time
which results in voltage.

The unit of a capacitor is farads, just as a resistor is ohms.
Common capacitor value terminology is farads, microfarads,
nanofarads, and picofarads.

1F   *1 farad*
1μF (microfarads)   *0.000001 (10^{-6}) farads*
1nF (nanofarad)   *0.000000001 (10^{-9}) farads*
1pF (picofarad)   *0.000000000001 (10^{-12}) farads*

At the instant (time t=0) when 1V is applied to the circuit in Figure
2.4-1, the capacitor is empty and has a voltage of 0V. Over
time, the capacitor is charged and the capacitor voltage starts
increasing. When the capacitor is full, the capacitor voltage is 1V.

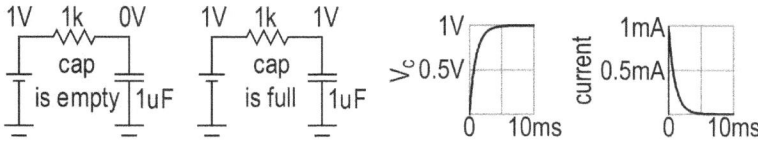

Figure 2.4-1

As the voltage increases, the current through the resistor
decreases

When V_c=0V   *I=(1V-0V)/1k=1mA*
When V_c=0.5V   *I=(1V-0.5V)/1k=0.5mA*
When V_c=1V   *I=(1V-1V)/1k=0mA*

The capacitor charge rate is the **time constant** (**T**). This indicates
when the capacitor reaches approximately (≈) 63.2% of the final
voltage.

R=1k, C=1μF
T=R*C
T=1k*1μF =1ms (millisecond)

Another useful equation is **rise time**. Rise time is defined as the time it takes to change from 10% of the initial voltage to 90% of the final voltage. Rise time is approximately 2.2 times the time constant.

$T=1ms$
$Rise\ time = \approx 2.2^*T = \approx 2.2ms$

2.5 Capacitor Parameters

The composition of capacitors is a form of ceramic, metal film, paper, mylar film, aluminum, tantalum, and electrolytic.
Values for capacitors are decade multiples of:

1.0 1.2 1.5 1.8 2.2 2.7 3.3 3.9 4.7 5.6 6.8 8.2

Common capacitor values are 2.2pF, 22pF, 220pF, 2.2nF, 22nF, 220nF, and 2.2µF.
It is common for ceramic capacitors to be 1µF or lower and non-ceramic capacitors to be 1µF or higher. Ceramic capacitors are smaller in physical size.
Capacitors tolerances can range from ±1% to ±20% or higher. They are categorized by temperature coefficients and temperature range of their dielectric materials.

C0G (NP0)	*±30ppm per °C (Class 1)*
X7R	*±15%, -55 to +125 °C (Class 2)*
X5R	*±15%, -55 to +85 °C (Class 2)*
Z5U	*+22% to -56%, +10 to +85 °C (Class 2)*
Y5V	*+22% to -82%, -30 to +85 °C (Class 2)*

The actual value of a Class 2 capacitor could be significantly less than the stated value, based on a combination of tolerance, temperature and applied voltage (both DC bias and high frequency AC).
A Y5V capacitor could yield less than 5% of the stated value (100µF Y5V might yield only 5µF of capacitance)
A capacitor can be destroyed when the voltage applied is too large. Capacitor voltage parameters are dependent on the composition. Applying AC voltage to a DC voltage rated component can also damage the capacitor and its dielectric.

Common ceramic capacitors voltages are 16V, 25V, 50V, 100V, 600V, 1000V and 2000V. Tantalum capacitors are rarely rated higher than 50V.

For industrial applications, where reliability is crucial, the capacitor voltage rating selected is twice that of the voltage applied. When the maximum voltage applied is 3.3V, the capacitor selected has a minimum rating of 6.6V. The practice of limiting the applied voltage lower than the rated voltage is referred to as **derating**.

The exception is the tantalum capacitor. For industrial applications, some Tantalum capacitors are recommended to be three times the voltage applied. For a 3.3V signal, the minimum rating is 9.9V.

For commercial applications, the capacitor voltage rating may be lower with a corresponding lower life expectancy. It is best to verify the recommended parameters in the datasheets.

Due to the chemistry of the electrolyte used in electrolytic capacitors, the life expectancy is affected by temperature, and follows the Arrhenius equation predictions. The result can be summarized as the life will be cut in half for every 10 °C increase in temperature.

The composition also affects polarization requirements. Non-ceramic capacitors, such as tantalum and electrolytic, are labeled with "+" and "-" symbols. The voltage applied to the "+" must be higher than the voltage applied to the "-". Ceramic capacitors do not have this restriction.

However, with the attempt to lower the cost of electronic components, it has been recorded that 1000V ceramic capacitors can fail when exposed to 707V AC.

ESR (equivalent series resistance) is also affected by the composition. Ceramic capacitors have a low ESR. This means they can handle high frequencies. Non-ceramic capacitors have a higher ESR. This means they cannot handle high frequencies.

ESR becomes increasingly important in switching power supply designs as switching frequencies continue to increase (see Voltage Regulators section).

There are a variety of capacitor physical sizes. Larger physical size capacitors can handle a higher voltage and can have a larger value.

Datasheets provide details about capacitor composition, physical size, values, and voltage ratings.

2.6 Parallel / Series Capacitors

A combination of parallel and series capacitors can be used when the required capacitor value is not standard. Series and parallel capacitor combinations are easy to calculate (Figures 2.6-1 and 2.6-2).

$$C_{parallel} = C1 + C2$$
$$C_{parallel} = 1\mu F + 1\mu F = 2\mu F$$

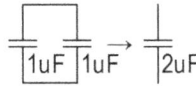

Figure 2.6-1

$$C_{series} = (C1*C2)/(C1+C2)$$
$$C_{series} = (1\mu F * 1\mu F)/(1\mu F + 1\mu F) = 0.5\mu F$$

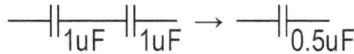

Figure 2.6-2

These are the exact opposite of series and parallel resistor calculations. However, the ESR of capacitors acts the same as resistors in series and parallel. You can achieve a much lower ESR by having ten 1μF capacitors in parallel than you could with one 10μF capacitor.

2.7 Capacitors (AC) / RC Filter

The world is filled with electrical noise. The AC characteristics of capacitors reduce electrical noise. Capacitors resist changes in voltage.

Capacitors are used on power inputs to ICs (integrated circuits) to reduce noise to the IC.

Low pass RC (resistor / capacitor) filters are common (Figure 2.7-1). Low pass means that lower frequencies are allowed and higher frequencies are attenuated (amplitudes are reduced).

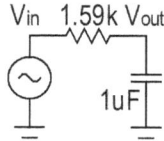

Figure 2.7-1

The frequency at which the RC filter starts attenuating is the **cutoff frequency** (corner frequency or break frequency). This is when the output voltage is 0.707 (1/√2) times the input voltage.

> $R=1.59k$ $C=1\mu f$ $\pi\ (pi)=3.14159$
> Cutoff frequency$=1/(2*\pi*R*C)$
> Cutoff frequency$=1/(2*\pi*1.59k*1\mu f)$
> Cutoff frequency$=100Hz$

With an input frequency of 10Hz in Figure 2.7-1, V_{out} is almost equal to V_{in}. At 100Hz, V_{out} is equal to approximately 70.7% of V_{in}. At 1kHz, V_{out} is equal to 10% of V_{in} (Figure 2.7-2).

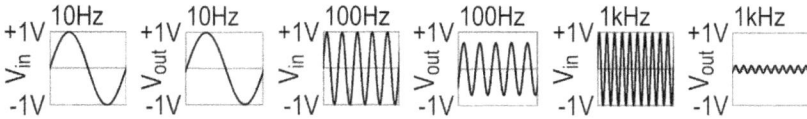

Figure 2.7-2

The cutoff frequency is also known as the **-3dB point**. The -3dB point is approximately one half of the passband power. The -6dB point is when the output is half of the input voltage.

Plotting the frequency on a log scale and the amplitude on a dB scale provides a good perspective (Figure 2.7-3).

> Plot equation: $y=20*log(x) \rightarrow x=10^{(y/20)}$
> when $y= -3dB \rightarrow x=10^{(-3/20)}$
> $x=0.707$ (70.7% of the input voltage)
> when $y= -20dB \rightarrow x=10^{(-10/20)}$
> $x=0.1$ (10% of the input voltage)

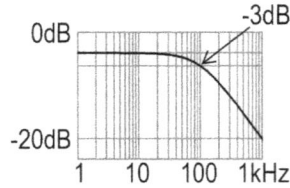
Figure 2.7-3

Phase shift is the amount of shift in the zero crossing of the output voltage relative to the input voltage. The output is delayed more at higher frequencies (Figure 2.7-4) because capacitors resist voltage changes.

1Hz	*no phase shift*
10Hz	*negligible phase shift*
100Hz	*45 degree phase shift*
1kHz	*approximately 90 degree phase shift*

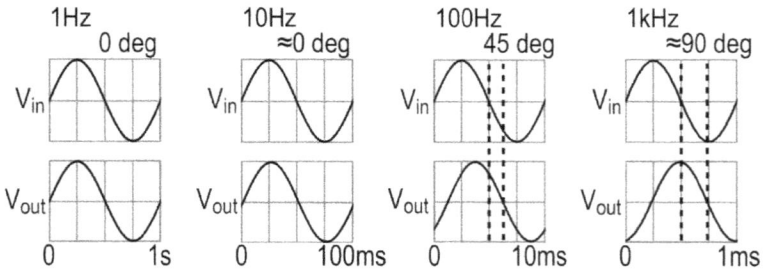
Figure 2.7-4

A phase plot displays the phase shift (Figure 2.7-5).

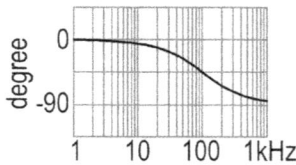
Figure 2.7-5

2.8 Cascade RC Filters

When a sharper RC filter is needed, cascading RC filters is an option (Figure 2.8-1).

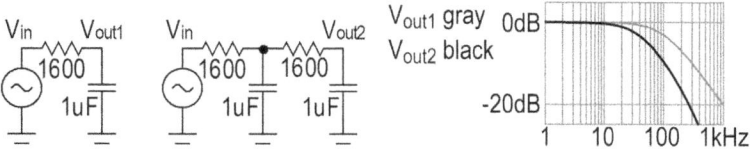

Figure 2.8-1

A sharper filter with minimal effect on the lower frequencies is to split the resistor, and add a second capacitor (Figure 2.8-2).

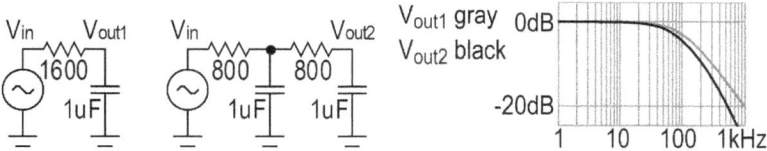

Figure 2.8-2

2.9 Inductors (DC)

An inductor is a coil of wire. Current through the wire creates a magnetic field. The magnetic field limits the rate of change of the electric current.

A sudden change in current is limited by the inductor. A change in current is allowed over a period of time.

The unit of an inductor is henries, just as a resistor is ohms. Common capacitor terminology is henries, millihenries and micro-henries.

1H	*1 henry*
1mH (millihenries)	*0.001 (10^{-3}) henries*
1µH (microhenries)	*0.000001 (10^{-6}) henries*

In Figure 2.9-1, At the instant (time t=0) when 1V is applied to the circuit, the current through the inductor is 0mA.

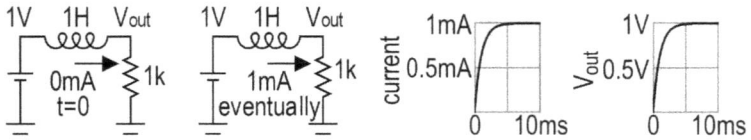

Figure 2.9-1

As current starts to flow through the inductor V_{out} voltage starts to increase. Eventually, V_{out} voltage equals the input voltage.

The inductor change rate is similar to the capacitor time constant (**T**). This indicates when the inductor reaches approximately 63.2% of the final current.

$R=1k, L=1H$
$T=L/R$
$T=1H/1k=1ms$

Many inductors contain a magnetic core comprised of iron or a ferrite material. The magnetic core circles the wire, or the wire is wrapped around the magnetic core (through the center). This increases the inductance.

2.10 Inductors (AC) / RL Filters

One use for inductors is to restrict the flow of AC current while allowing DC current to pass. The AC characteristics of an inductor provide a way to reduce electronic noise. Inductors designed for this use are also known as chokes.

Chokes can be used to separate frequencies. Used with capacitors, they allow tuning to a specific frequency. This has been used for radios and TVs.

One of the most common inductor filters is an inductor in series with the input voltage / current (Figure 2.10-1). This attenuates high frequency noise.

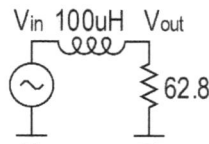

Figure 2.10-1

A useful equation for determining the frequency at which an RL filter starts attenuating is the cutoff frequency (corner frequency or break frequency). This is when the output is 0.707 (1/√2) times the input.

R=62.8, L=100µH, π (pi)=3.14159
Cutoff frequency=R/(2*π*L)
Cutoff frequency= 62.8/(2*π*100µH)=100kHz

Using Figure 2.10-1, with an input of 10kHz, the V_{out} is almost equal to the V_{in}. At 100kHz, V_{out} is equal to approximately 70.7% of V_{in}. At 1MHz, V_{out} is equal to 10% of V_{in} (Figure 2.10-2)

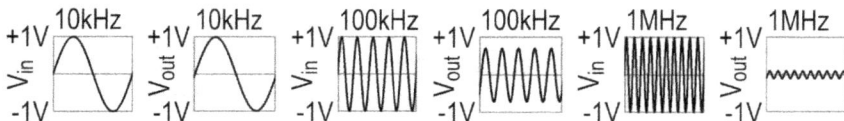

Figure 2.10-2

Phase shift is the amount of shift in the zero crossing of the current relative to the input voltage. The current is delayed more at higher frequencies (Figure 2.10-3) because inductors resist current changes.

1kHz   no phase shift
10kHz   negligible phase shift
100kHz   45 degree phase shift
1MHz   approximately 90 degree phase shift

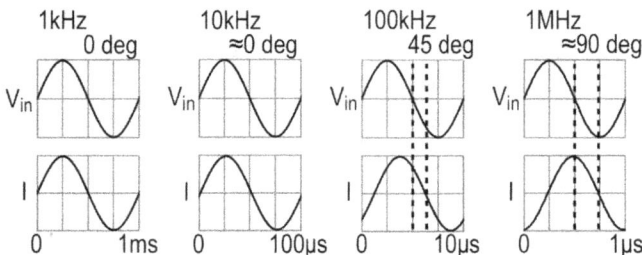

Figure 2.10-3

2.11 Impedance

Impedance is the resistance to the flow of alternating current (AC). The word impedance can be used for resistors, capacitors, and inductors usually when they are combined. The letter "Z" is used to denote impedance.

Impedance has a direct relationship to frequency as described in the capacitor and inductor sections.

2.12 Crystals

Crystals (crystal oscillators) generate a constant, precise frequency. Precise frequencies are needed for wireless communication and for some wired communication. They are also used by micros (microcontrollers / microprocessors).

It is common for crystals to range from kilohertz to hundreds of megahertz. Items, such as watches, use crystals to ensure that the time displayed is consistent and accurate.

Crystals are unable to provide much current. This restriction requires connecting to a high input impedance component which draws microamps (or picoamps) of current.

Small value capacitors (between 5pF and 22pF) are used on both crystal outputs. The crystal and the capacitors are commonly placed in close proximity to the micro (Figure 2.12-1).

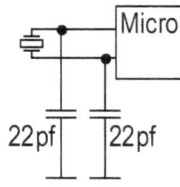

Figure 2.12-1

Crystal frequencies can change a small amount over temperature. Datasheets describe these parameters.

Lower cost components are resonators. Resonators provide a fairly constant frequency, but change more than crystals over temperature.

A third category is oscillators. These components are able to provide current, sometimes up to 16mA. They are used when multiple devices need a constant frequency.

2.13 Fuses

A fuse disrupts current flow when a maximum current is reached. When the maximum current is exceeded, the fuse breaks an internal connection, usually by melting a conductor.

Low current fuses, in the low mA range, may require special treatment when soldered. Too much heat during the soldering process could damage the fuse.

While SMT fuses are smaller and easier to install, the very low current fuses are available only in a PTH configuration. The low current PTH fuses often use a socket into which to insert the fuse after the socket has been soldered to the PCB.

A resettable fuse works similar to a regular fuse, except it is self-healing. Excessive current often creates heat. This heat in a resettable fuse causes the resistance to increase significantly. This causes a disruption in current flow. After the temperature cools, the resettable fuse returns to normal operation.

2.14 Transformers

A transformer is a combination of two inductors. AC current flowing through one inductor (primary) creates a magnetic field. A second inductor (secondary), closely coupled to the first inductor, converts the magnetic field it detects back into current. Both inductors are usually wrapped around the same magnetic core.

Transformers are comprised of galvanic isolation. The primary and secondary are completely isolated with no current flowing between the primary coil and the secondary coil.

A useful aspect is the ability to increase or decrease the current by increasing or decreasing the number of times the wire wraps around the magnetic core for both inductors. When the primary has twice as many turns of wire as the secondary, the output current of the secondary is doubled and the voltage is cut in half.

The common nomenclature is to specify transformers in terms of voltage instead of current. A 2 to 1 transformer would convert 10V AC into 5V AC. Because there are losses in transformers, a 2 to 1 transformer would convert 10V AC into something less than 5V AC.

Transformers are used in many ways. One of the first usages was for transmission of electric power.

Wire has resistance. Resistance to current flow results in loss of power. For example, common electric wire used for a house is 12 gauge. A one foot long piece of 12 gauge wire has approximately 0.001588Ω resistance.

For short distances, this is not a problem. However for long distances, with many miles, resistance does become a problem. Longer wire → higher resistance → more power → higher heat.

Using a larger size (diameter) wire reduces the resistance. However, there is a limit. At some point the wire is too heavy to hang on towers.

The use of transformers for electric power transmission reduces this problem by increasing the voltage, but more importantly decreasing the current in the transmission lines. The power loss is proportional to the resistance of the wire and to the current squared. Reducing the current results in a square root reduction in power loss, $P=I^{2*}R$ (see the Parallel / Series Resistors section).

Electric power generation stations use transformers to increase the voltage and decrease the current. The high voltage / low current electricity is transmitted long distances. When the electricity reaches the destination, transformers decrease the voltage and increase the current for our use.

It is not unusual for modern electricity transmission lines to use 110,000V for electricity transmission at 60Hz. Much of the world may use other voltages at 50Hz.

One exception is Japan which has both 50Hz and 60Hz electricity. This frequency difference can make power sharing challenging from one part of the country to another part of the country.

At the destination, the voltage is converted to lower voltage. The common voltage for most U.S. homes and businesses is 110V. For other countries, the common voltage is 220V.

Most electronic equipment uses transformers to reduce the AC voltage **and** convert it into a DC voltage.

The Diode Rectifier section describes the conversion of AC voltages to DC voltages.

2.15 Common Mode Chokes

Common mode chokes are two inductors combined into one component. Two wires are wound in parallel on a single magnetic core forming a highly coupled 1:1 transformer.

This presents high impedances to block common mode currents (undesired noisy currents) and low impedances to normal current flow.

Common mode chokes allow I_{in} to flow when it equals I_{out}. I_{in} that does not equal I_{out} is blocked (Figure 2.15-1).

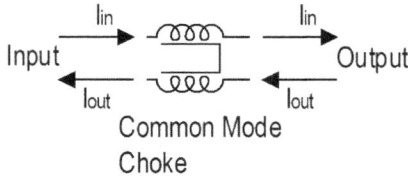

Common Mode
Choke
Figure 2.15-1

Common mode chokes reduce external electronic noise from entering the electronic assembly. They also reduce electronic assembly noise from exiting the assembly which could affect other electronic equipment. This can be crucial for meeting EMC requirements.

Placing a common mode choke between one electronic assembly voltage / ground section and a second voltage / ground section can help keep noise on one section from affecting the other section. This could be useful for electronics that control electric motors which generate electronic noise.

2.16 Hall Sensors

Hall Sensors are unique. They detect magnetic fields while stationary.

The Electro-Magnetism section describes how swinging a magnet past a wire generates current. The Hall sensor does not require movement. It detects magnets with nothing moving.

Hall sensors have many uses. One of the most common uses is motors. Magnets on a motor shaft, coupled with Hall sensors, indicate to the electronics the position of the motor windings. This allows the electronics to energize the correct motor windings at the correct time.

Chapter 2 Quiz

R	resistor	V	voltage
C	capacitor	I	current
L	inductor	T	time constant

2-1   What resistor package size is required for a voltage of 1V across a 100Ω resistor?

2-2   With a 1206 package size, what is the maximum current through a 2.49k resistor?

2-3 The voltage across two resistors in series is 2.5V. The sum of the two resistors is 5k. The required voltage across the second resistor is 1V. What is the second resistor value?

2-4 Two resistors are in parallel. One resistor value is 10k. The combined resistance of the parallel resistors is 2k. What is the second resistor value?

2-5 A low pass RC filter with C=10μF and T=50ms; what is R?

2-6 A low pass RC filter with R=20k and C=10nF; what is the rise time?

2-7 A low pass RC filter with R=200k and C=0.1μF; what is the cutoff frequency?

2-8 A low pass RC filter with R=15k and C=47nF has a 1V input. The output is 0.1V. What is the input frequency?

2-9 What are three advantages of ceramic capacitors vs. non-ceramic capacitors?

2-10 With an applied voltage of 7V, what is the minimum voltage rating for a tantalum capacitor in an industrial product?

2-11 Two capacitors are in parallel. One capacitor value is 4.7μF. The combined capacitance of the parallel capacitors is 11.5μF. What is the second capacitor value?

2-12 Two capacitors are in series. One capacitor value is 27nF. The combined capacitance of the series capacitors is 14.85nF. What is second series capacitor value?

2-13 A low pass RL filter with T=8μs and R=17.4k; what is the value of the inductor?

2-14 What common material is used to increase the magnetic field for inductors?

2-15 A low pass RL filter with R=10Ω and L= 22μH; what is the -3dB frequency?

2-16 A low pass RL filter with R=2.67Ω and L=10μH has a 3V input. The output is 2.121V. What is the frequency of the input?

2-17 What three types of electronic components provide a constant frequency output?

2-18 What common parameter causes component characteristics to change?

2-19 During the electronic manufacturing process, what can be a problem with low current SMT fuses?

2-20 What eliminates the potential problem of replacing fuses after a product has been released to customers?

2-21   What can be done to reduce the loss of electric power (electricity) over long distances?

2-22   What are the common frequencies used for electricity throughout the world?

2-23   What is unique about Hall sensors?

2-24   What is one of the most common uses for Hall sensors?

Chapter 2 Design Challenges (use SPICE models when applicable)

2:1   An analog input signal could be as high as 7V. The receiving component maximum input is 2.5V with an input impedance of 100k.

  Design a resistor divider for a maximum voltage input to the receiving component that does not exceed 2.5V. Allow for 1% tolerances.

2:2   An analog input signal ranges from 0 to 2V. The input frequency can vary from 0 to 1kHz. There is high frequency electronic noise on the input. The receiving electronics has a high input impedance.

  Design an RC filter (two resistors and two capacitors) that reduces the electronic noise without affecting the input signal. Use a SPICE model to help identify a good solution.

2:3   A power rail on an electronic assembly has an input impedance of 50Ω. The electronic assembly radiates via the power rail 100MHz.

  Determine the optimum inductor size to use to block the 100MHz from leaving the electronic assembly.

2:4   A signal requires 0.1µF capacitance that is capable of handling 1000V spikes.

  The capacitors in this application are a maximum of 1000V. Determine a capacitor combination that will meet this requirement.

Chapter 3

ANALOG DISCRETE ACTIVE COMPONENTS

3.1 Diodes

Diodes allow current to flow only in one direction. Current flows into the anode and out of the cathode (Figure 3.1-1). The voltage is reduced by approximately 0.6V (forward voltage drop).

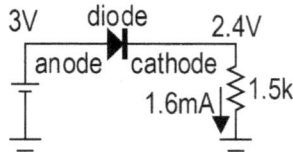

Figure 3.1-1

Forward voltage drops increase with larger currents. Datasheets specify the forward voltage drop as a function of current.

A diode connected backwards blocks most of current flow. Leakage current is the small amount of current that does flow (Figure 3.1-2).

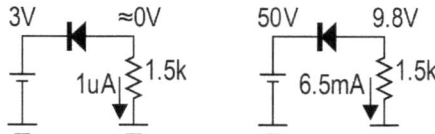

Figure 3.1-2

Larger voltages, to a diode connected backwards, does allow current to flow through the diode (Figure 3.1-2).

3.2 LEDs

LEDs are light emitting diodes. As this is also a diode, it allows current to flow only in one direction.

When current flows through the LED, it illuminates. It is common to use approximately 10mA of current to illuminate an LED.

LEDs are reliable. They last for many years. Exceptions are new technology LED lights which are intended to replace incandescent light bulbs. These LED lights are clustered close to each other and use extra current which can cause them to generate heat and eventually fail.

When LEDs are used simply as an indicator, not light illumination, they are exceptional. Even illuminating an LED for as little as 10ms can generate enough light for the eye to easily detect.

3.3 Diode Rectifiers

Diode rectifiers are used with AC voltages. A halfwave rectifier blocks half of the voltage because current is allowed to flow only in one direction (Figure 3.3-1).

Figure 3.3-1

The output is approximately 0.6V smaller than the input. This is due to the diode voltage drop.

A fullwave rectifier allows both halves of an AC voltage to pass. The output is approximately 1.2V smaller than the input. This is due to two diode voltage drops (Figure 3.3-2).

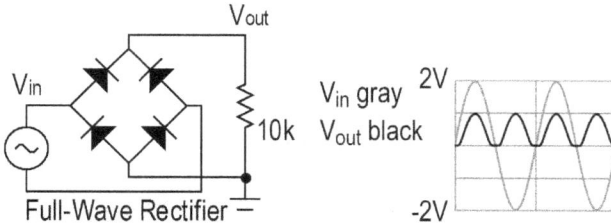

Figure 3.3-2

Adding a large value capacitor smooths the output voltage. The output starts looking similar to a DC voltage (Figure 3.3-3).

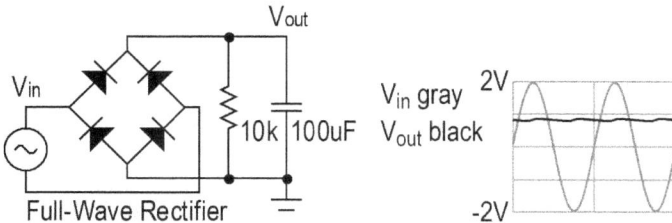

Figure 3.3-3

3.4 Zener Diodes

Zener diodes are designed to provide an accurate reverse voltage drop. In the forward direction, they behave like diodes. In the reverse direction, they limit the current until the Zener knee voltage is reached.

A 5.1V BZT52C5V1 Zener diode that has a knee voltage at 5.1V. This Zener diode limits the voltage to approximately 5.1V. As the current through the Zener diode increases, the knee voltage increases a small amount (Figures 3.4-1 and 3.4-2).

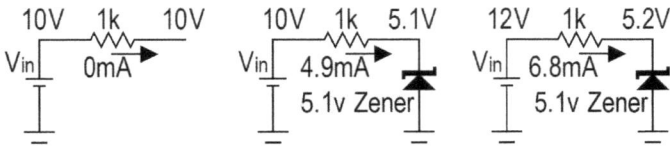

Figure 3.4-1

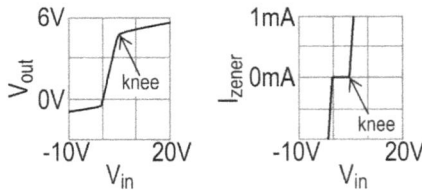

Figure 3.4-2

The Zener works like a diode when the voltage reaches approximately 0.6V.

Different Zener diodes have different characteristics. The 1N5338B is also a 5.1V Zener. The Zener voltage changes less with increased current than the BZT52C5V1 (Figure 3.4-3).

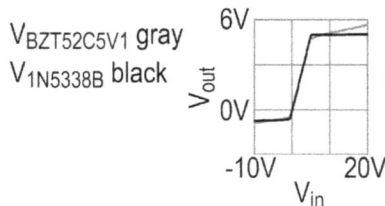

$V_{BZT52C5V1}$ gray
$V_{1N5338B}$ black

Figure 3.4-3

Zener diode datasheets show the knee voltage occurring at a negative voltage (Figure 3.4-4). This is because input voltage is applied to the anode end of the diode.

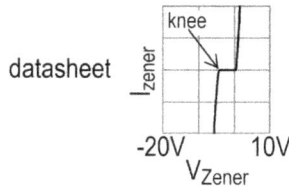

Figure 3.4-4

3.5 FETs (Field Effect Transistors)

A transistor is similar to a water faucet. It controls current as a water faucet controls water.

There are three types of transistors; FETs (MOSFET), JFETs, and BJTs (bipolar junction transistors). FETs and JFETs are controlled by voltage (the gate current is commonly in the nanoamp range). BJTs are controlled by current.

A FET is either a P Channel or an N Channel (Figure 3.5-1). A P Channel is "on" with a gate voltage lower than the source voltage. An N Channel is "on" with a gate voltage higher than the source voltage (with a variation in the depletion mode).

Figure 3.5-1

P Channels are "on" until the gate voltage is larger than the source voltage minus the gate threshold voltage. BSS84LT FETs have a gate threshold voltage of approximately 2V (Figure 3.5-2).

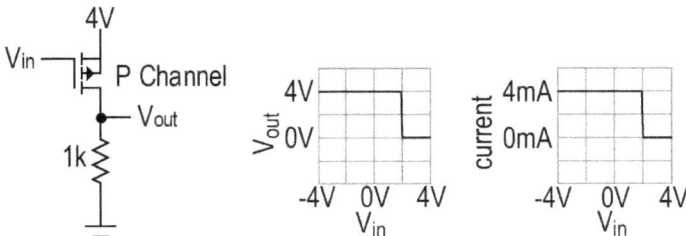

Figure 3.5-2

The N Channel FET is either an enhancement mode FET or depletion mode FET.

N Channel enhancement FETs are "off" until the gate voltage is larger than the source voltage plus the gate threshold voltage. BSS138 FETs have a gate threshold voltage of approximately 1.5V (Figure 3.5-3).

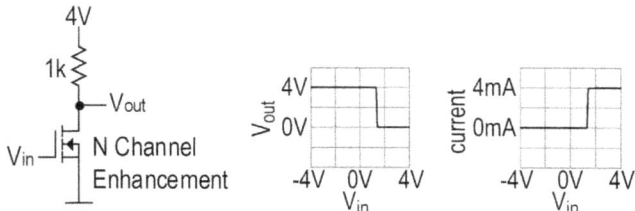

Figure 3.5-3

N Channel depletion FETs are "off" until the gate voltage is larger than the source voltage minus the gate threshold voltage. BSS159N FETs have a gate threshold voltage of approximately 2V (Figure 3.5-4).

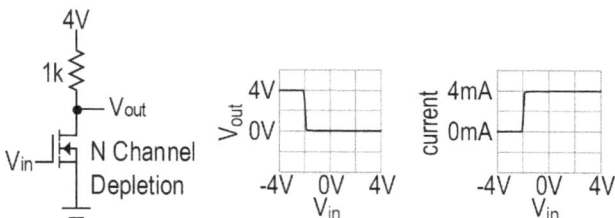

Figure 3.5-4

Source voltages are rarely higher than drain voltages. When the source voltage exceeds the drain voltage by approximately 0.6V (a diode forward voltage drop), the current flows backwards through the FET. Inserting a diode in series with the FET source eliminates this problem (Figure 3.5-5).

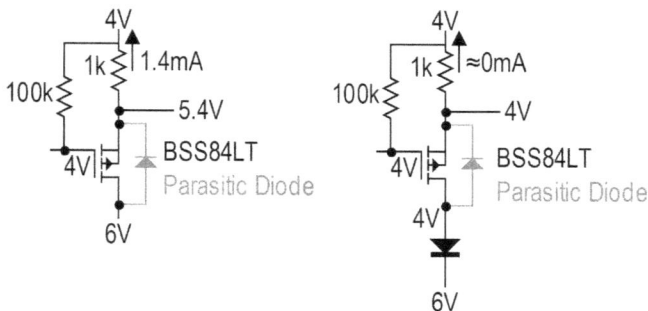

Figure 3.5-5

One consideration is to ensure the FET is "off" when power is first applied to an assembly. Most micro (microcontroller / microprocessor) pins are set to be an input when power is first applied. When power is first applied to an assembly, there are no output voltages from the micro.

Pull-ups on P Channel gates and pull-downs on N Channels gates ensures the FETs are "off" during power-up (Figure 3.5-6).

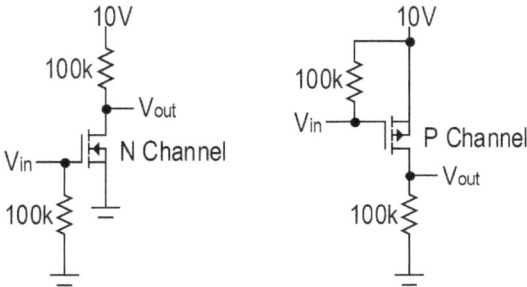

Figure 3.5-6

3.6 JFETs (Junction Field Effect Transistors)

While a JFET is similar to a FET, there are some distinct differences.

The JFET input impedance is not as large as the FET input impedance. This results in larger gate current to control the JFET.

JFETs operate in a depletion mode only. The FET operates in a depletion mode or enhancement mode.

JFETs have a higher drain resistance.

FETs are more commonly used in electronic assemblies.

3.7 BJTs (Bipolar Junction Transistors)

A BJT is similar to a FET. The primary difference is that a BJT is controlled by current vs. a FET which is controlled by voltage. A BJT is either a PNP or an NPN (Figure 3.7-1).

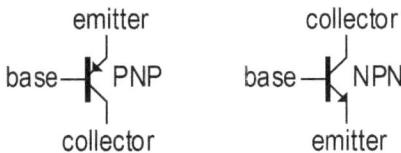

Figure 3.7-1

For a PNP, as current flows out of the base, current flows into the emitter and out of the collector (Figure 3.7-2).

$$I_{collector} = I_{base} * gain$$
$$I_{emitter} = I_{collector} + I_{base}$$

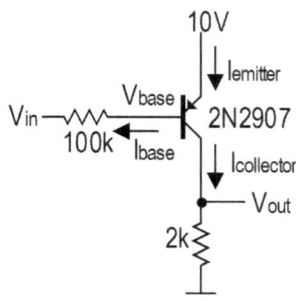

Figure 3.7-2

For a PNP, V_{base} must be smaller than $V_{emitter}$ by approximately 0.6V, before the BJT conducts collector current. After the appropriate base voltage (V_{base}) is reached, the base current (I_{base}) starts to flow out of the base. The emitter current and collector current follow the base current, but are amplified (Figure 3.7-3).

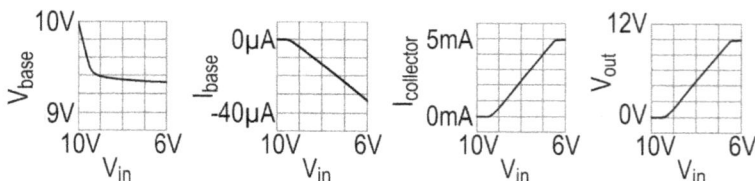

Figure 3.7-3

The gain is approximately 170 for a 2N2907. The collector current is approximately 170 times larger than the base current (Figure 3.7-3).

The limiting factor, in Figure 3.7-2, is the 2k resistor. After the collector current reaches 5mA, the voltage across the 2k resistor is 10V. This restricts the maximum collector current to 5mA, even when the base current increases.

For an NPN, as current flows into the base, current flows into the collector and out of the emitter (Figure 3.7-4).

$$I_{collector} = I_{base} * gain$$
$$I_{emitter} = I_{collector} + I_{base}$$

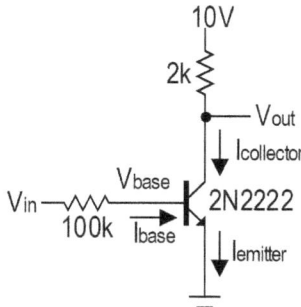

Figure 3.7-4

For an NPN, V_{base} must exceed $V_{emitter}$ by approximately 0.6V. After the appropriate base voltage (V_{base}) is reached, the base current (I_{base}) starts to flow into the base. The emitter current and collector current follow the base current, but are amplified (Figure 3.7-5).

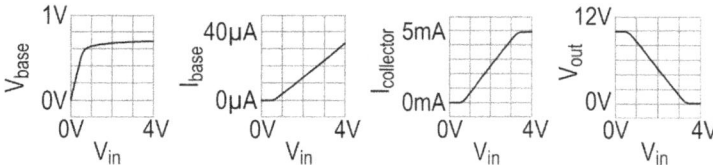

Figure 3.7-5

The gain is approximately 200 for a 2N2222. The collector current is approximately 200 times larger than the base current (Figure 3.7-5).

The limiting factor in Figure 3.7-4 is the 2k resistor. After the collector current reaches 5mA, the voltage across the 2k resistor is 10V. This restricts the maximum collector current to 5mA, even when the base current increases.

3.8 Analog Switch

An analog switch is like a regular switch, except it is controlled by voltage. The Maxim MAX4730 has both NO (normally open) and NC (normally closed) input / output (Figure 3.8-1).

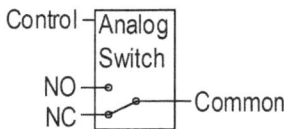

Figure 3.8-1

Parameters to consider include R_{on} (resistance between input and output) and bandwidth (maximum speed of input / output). Data-sheets describe other parameters which could impact the design.

An analog switch is different than a digital switch. When in the "on" mode, an analog switch is like a low value resistor. Signal "out" equals signal "in" unless R_{on} attenuates the signal.

3.9 Digital Potentiometer

A digital potentiometer provides the ability to electronically control a resistor value (Figure 3.9-1).

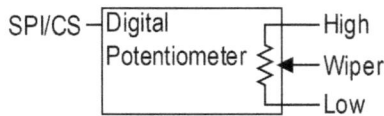

Figure 3.9-1

The Maxim MAX5401 provides a maximum of 100k (resistance between High and Low). It divides this resistance into 255 increments (256 values).

> 0 *0 ohms (100k/255*0) between Low and Wiper*
> 1 *≈ 392.2 ohms (100k/255*1) between Low and Wiper*
> 2 *≈ 784.3 ohms (100k/255*2) between Low and Wiper*
> 3 *≈ 1176.5 ohms (100k/255*3) between Low and Wiper*
> ...
> 255 *100k ohms (100k/255*255) between Low and Wiper*

The resistor value is controlled by SPI communication (see the SPI Serial Communication section)

3.10 Optoisolators

Optoisolators (optocoupler or photocoupler) contain an LED and a phototransistor. The LED light is detected by a phototransistor. The brighter the LED, the more current in the phototransistor.

Optoisolators provide galvanic isolation. The input voltage / current are not electrically connected to the output voltage / current. The input current does affect the output current (Figure 3.10-1).

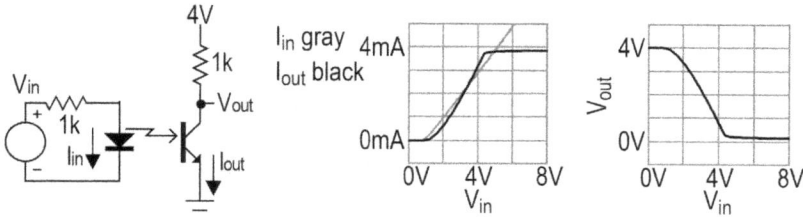

Figure 3.10-1

The input current does not begin to flow until the input voltage exceeds a diode voltage drop, which is approximately 1.6V. As the input current increases, the output current increases. The output current is limited by the 1k resistor.

Commercially available optoisolators can withstand input voltage to output voltage differences of up to 10,000V.

Chapter 3 Quiz

3-1   The voltage across a resistor and Schottky diode in series is 3.3V. The diode forward voltage drop is 0.5V. The current through the diode is 2mA. What is the value of the resistor?

3-2   The voltage across a 1k resistor and a Schottky diode in series is 3V. Why is the current through the diode only 1μA?

3-3   The voltage across a 576Ω resistor and Zener diode in series is 12V. The current through the Zener diode is 5mA. What is the approximate value of the Zener diode knee voltage?

3-4   What happens to the knee voltage when the current through a Zener diode increases?

3-5   In what direction does the gate current flow in an N Channel FET?

3-6   The gate threshold voltage for a P Channel FET is 2.1V. The source voltage is 5V. At what gate voltage does the FET start conducting current?

3-7   What is the primary difference between a BJT and an FET?

3-8   What is the effect of gain in a BJT?

3-9   An optoisolator usually is comprised of what two components?

Chapter 3 Design Challenges (use SPICE models when applicable)

3:1   An input signal ranges from 0V to 24V. This voltage controls a BSS138 N Channel FET. The BSS138 maximum gate to source voltage is ±20V.

Design a circuit that will protect the BSS138.

3:2   An electronic assembly has an LED. There is a 3V signal from a micro that controls the LED.

Design a circuit that provides a micro to LED interface.

3:3   A double throw switch controls an electronic assembly. The pullup resistors to the switch contacts are connected to 12V.

Design a circuit that provides an appropriate interface to a micro powered by 3V using FETs. The switch debounce needs to be handled by the hardware, not by the micro code.

Chapter 4

ANALOG MICROCIRCUIT ACTIVE COMPONENTS

4.1 Comparators

A comparator compares voltages. When the "+" input voltage is smaller than the "-" input voltage, the output voltage equals the negative power supply voltage. When it is larger, the output voltage equals the positive power supply voltage (Figure 4.1-1).

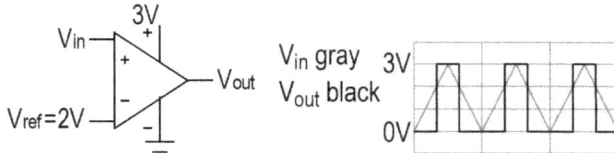

Figure 4.1-1

While some ICs can be powered by a positive voltage and a negative voltage, it is common to power ICs with a positive voltage and ground.

Sometimes the input voltage is noisy. This can cause a problem when the input voltage is close to the reference voltage (Figure 4.1-2).

Figure 4.1-2

Noisy input voltages can trigger multiple output voltage pulses inadvertently. A good way to reduce the problem is to incorporate hysteresis. Hysteresis is positive feedback which changes one of the input voltages a small amount.

The resistor configuration and values chosen in Figure 4.1-3 eliminate the inadvertent output voltage pulses. The comparator "+" and "-" inputs have a high resistance (very little current flows into those inputs).

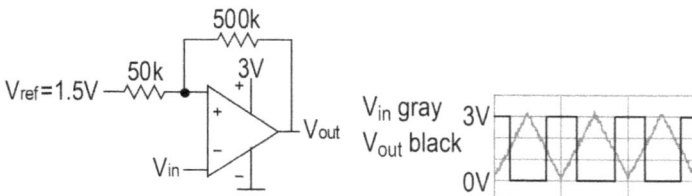

Figure 4.1-3

Hysteresis changes the trip point voltages. While the op amp configuration is set to trip at 1.5V the point at which the output voltage changes is no longer at 1.5V (Figure 4.1-4).

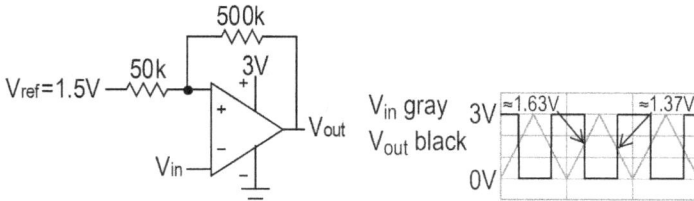

Figure 4.1-4

One option for comparators is an open collector output. When the "+" input voltage is larger than the "-" input voltage, the output voltage is floating instead of the power supply voltage rail. In Figure 4.1-5, a pullup resistor to 1.8V results in an output voltage that is either 0V or 1.8V.

Figure 4.1-5

Open collector outputs provide the option of changing the voltage output to something other than the supply voltage and / or connecting the output of multiple comparators to each other.

4.2 Op Amps

It is common for op amps to amplify a voltage. Op amps can be used as a comparator, but comparators are faster.

An op amp is at equilibrium when the "-" input voltage equals the "+" input voltage. The output voltage will not change (Figure 4.2-1).

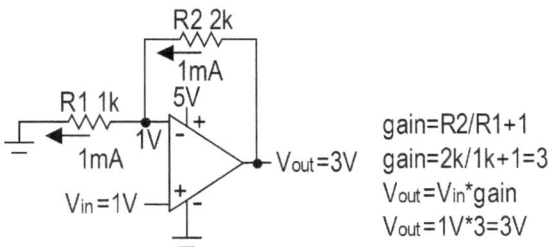

gain=R2/R1+1
gain=2k/1k+1=3
$V_{out}=V_{in}*gain$
$V_{out}=1V*3=3V$

Figure 4.2-1

When power is initially applied to an op amp, the "-" input is 0V, the "+" input is 1V, and the output is 0V. When the "+" input voltage is larger than the "-" input voltage, the output voltage starts to increase.

As the output voltage increases, "-" input voltage increases (resistor voltage divider). When the "-" input voltage equals the "+" input voltage, the output voltage stops changing.

The "+" and "-" inputs have a high resistance (very little current flows into those inputs). In Figure 4.2-1, the current through R1 and R2 is 1mA. This resistor configuration has a gain of 3.

Connecting the output to the "-" input results in the output voltage equaling the input voltage (Figure 4.2-2).

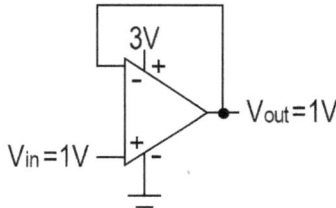

Figure 4.2-2

A negative gain can also be achieved (Figure 4.2-3).

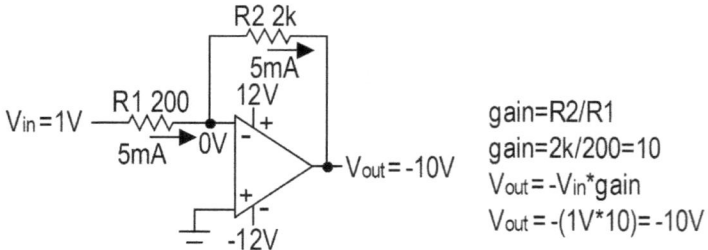

$$gain=R2/R1$$
$$gain=2k/200=10$$
$$V_{out}=-V_{in}*gain$$
$$V_{out}=-(1V*10)=-10V$$

Figure 4.2-3

Op amp parameters that affect the output include:

- *Input Offset Voltage is the voltage difference between the "+" input and "-" input for the output to not change. For a perfect op amp, that difference is 0V. In real life, that difference can be in the range of 1mV or more. This becomes increasingly important for small input voltages and large gains. This offset voltage can add error to the output voltage.*

- *Input Bias Current is the amount of current into the "+" input and amount of current into the "-" input. For a perfect op amp, the input bias current is 0A. In real life, that current can be in the range of 500pA or more. This is rarely a problem, unless high value resistors in the megaohm range are used.*
- *Bandwidth is the input voltage maximum frequency which the output voltage can follow.*
- *Slew rate is the maximum speed the output voltage can change.*

There are trade-offs with op amps. Low power op amps are slower.

4.3 Op Amp Offset

An op amp provides a voltage output relative to ground. Sometimes the voltage output needs to be shifted relative to ground.

Adding a bias voltage will shift the voltage output. It will not affect the gain (Figures 4.3-1 and 4.3-2).

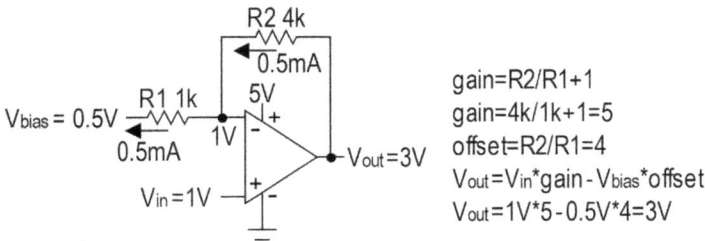

gain=R2/R1+1
gain=4k/1k+1=5
offset=R2/R1=4
$V_{out}=V_{in}*gain-V_{bias}*offset$
$V_{out}=1V*5-0.5V*4=3V$

Figure 4.3-1

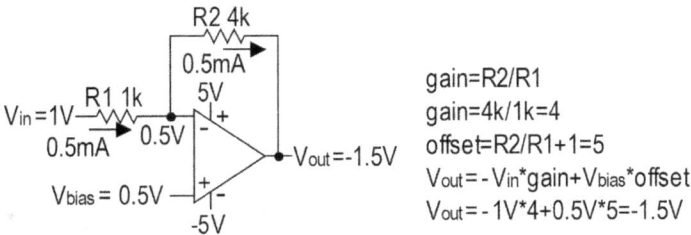

gain=R2/R1
gain=4k/1k=4
offset=R2/R1+1=5
$V_{out}=-V_{in}*gain+V_{bias}*offset$
$V_{out}=-1V*4+0.5V*5=-1.5V$

Figure 4.3-2

4.4 Variable Gain / Variable Offset Op Amp

A variable gain op amp output and variable offset op amp output are achievable with digital potentiometers (Figure 4.4-1).

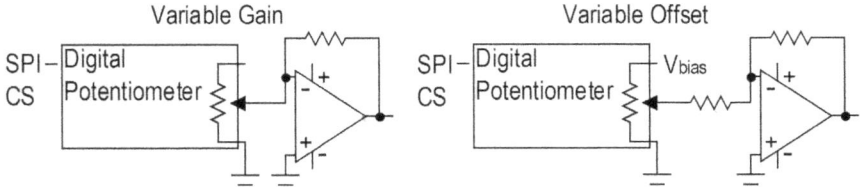

Figure 4.4-1

4.5 Differential Op Amp

A differential op amp configuration can work well when a signal is available as a differential voltage (Figure 4.5-1).

The input signal can be amplified with reduced amplification of noise. The noise amplification can be reduced more by using higher precision resistors. If this is an inadequate reduction in noise, an instrumentation amp may be required.

A disadvantage of a differential op amp is the input resistance. Current flowing into R1 and R3 could affect the input voltage.

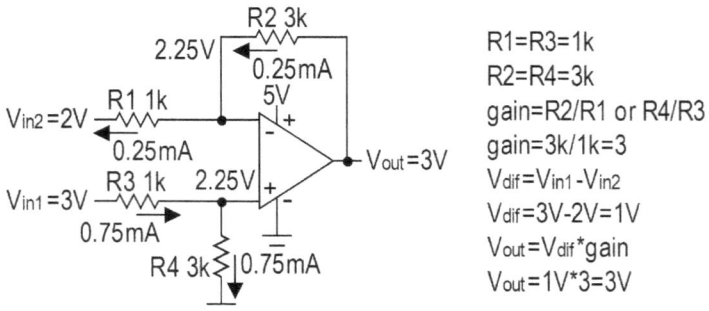

Figure 4.5-1

4.6 Differential Input / Output Op Amp

Electronic noise is often a problem, especially when traveling a long distance. Referencing signals to ground can work, but sometimes the ground is also noisy.

The best way to reduce noise is to use a differential input / output configuration (Figure 4.6-1). This references a signal between two outputs rather than referencing a signal between an output and ground.

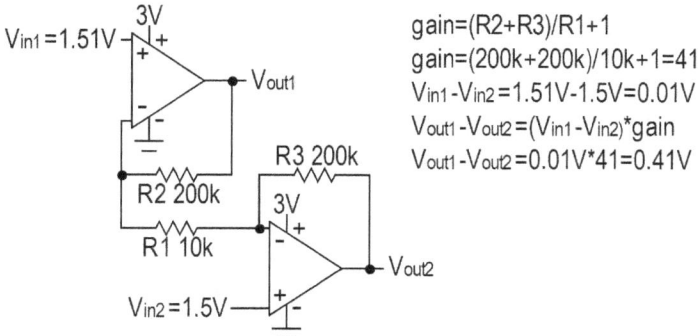

$$gain=(R2+R3)/R1+1$$
$$gain=(200k+200k)/10k+1=41$$
$$V_{in1}-V_{in2}=1.51V-1.5V=0.01V$$
$$V_{out1}-V_{out2}=(V_{in1}-V_{in2})*gain$$
$$V_{out1}-V_{out2}=0.01V*41=0.41V$$

Figure 4.6-1

The configuration in Figure 4.6-1 amplifies the input signal but does not amplify the common mode noise. The common mode noise passes at unity gain. Using higher precision resistors for R2 and R3 would reduce the noise more.

A differential input / output op amp configuration may not be linear if the inputs approach either power rail.

Instrumentation amps are a good way to convert differential voltages to a single ended voltage on the receiving end.

4.7 Differential Input / Output Op Amp Filtering

Sometimes a sensor detects electronic noise. Filtering a differential input / output signal can help (Figure 4.7-1).

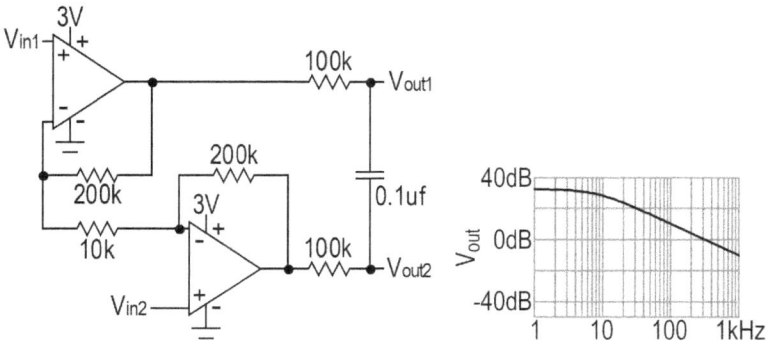

Figure 4.7-1

A 200k resistor with a 0.1μF capacitor has a 7.96Hz cutoff frequency. For this differential input / output op amp configuration, splitting the 200k resistor into two 100k output resistors is necessary for a differential configuration.

Cutoff frequency=1/(2*π*R*C)
Cutoff frequency=1/(2*π*200k*0.1μf)=7.96Hz

Another filter option is to place 0.1μF capacitors across the 200k feedback resistors (Figure 4.7-2). This provides additional filtering, up to a point. At high frequencies, the filtering stops as the op amp approaches unity gain.

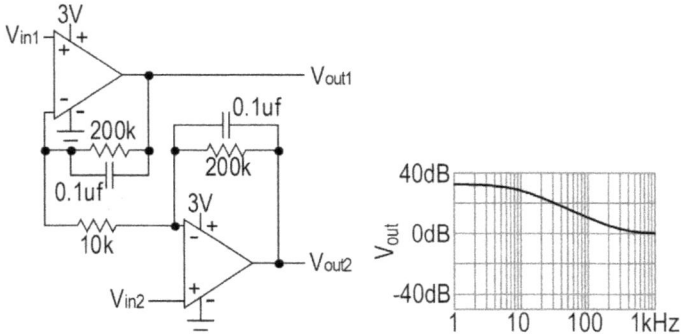

Figure 4.7-2

Combining the two filters results in a sharper filter (Figure 4.7-3).

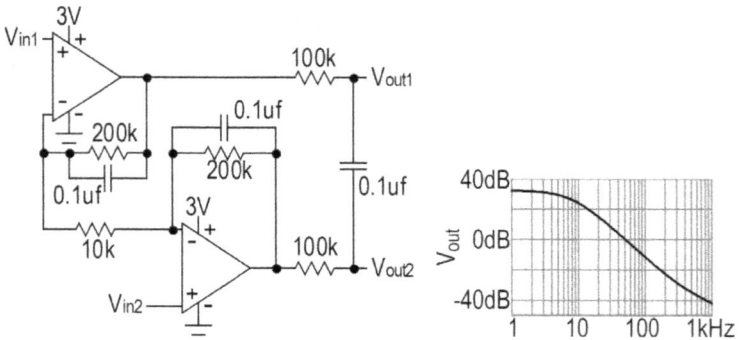

Figure 4.7-3

4.8 Rail to Rail Op Amp

Op amp output voltages can become non-linear as the outputs approach the power rail. Rail-to-rail op amps allow output voltages to approach the voltage rails with minimal effect.

Some op amps allow rail-to-rail inputs. This means the input voltages can approach the voltage rails with minimal effect.

A rail-to-rail description may not be completely accurate. Rail-to-rail usually means to within a few hundred mV or less. Datasheets provide these parameters.

4.9 Sensors / Constant Current / Voltage

Some sensors are a resistor bridge. Pressure sensors are one example. As the pressure changes, the resistance changes.

In Figure 4.9-1, the output is a differential output. Two different pressures result in two different differential voltage outputs.

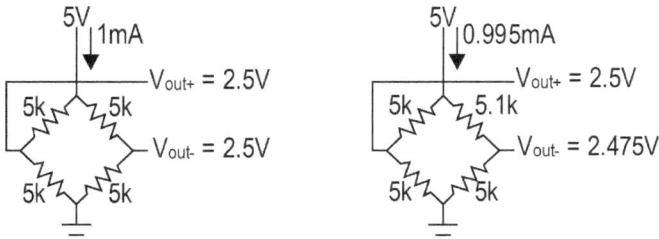

Figure 4.9-1

For a voltage across a sensor, changes in resistance affect the current through the sensor. Using a constant current source for sensors in series eliminates this problem (Figure 4.9-2).

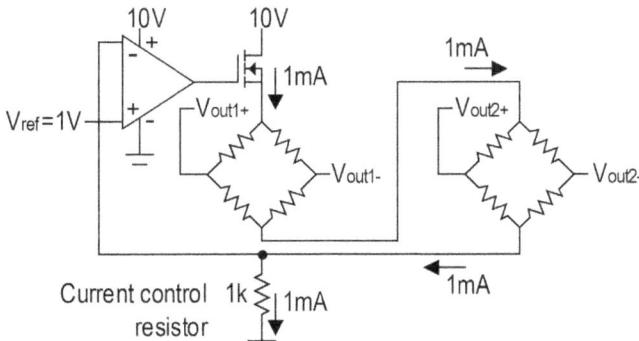

Figure 4.9-2

Low power op amps may have limited output current. A FET could provide the extra current.

To reduce costs, a constant reference voltage could be used with parallel sensors (Figure 4.9-3). Changes in sensor resistance should not affect the performance of the other sensors.

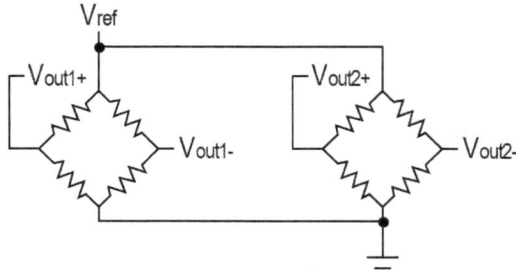

Figure 4.9-3

4.10 Instrumentations Amps

Electronic noise is common. Amplifying a signal can amplify noise. Instrumentation amps minimize noise (Figure 4.10-1).

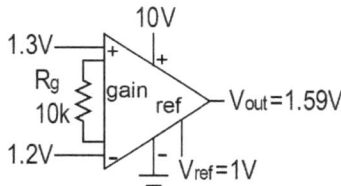

INA826
gain=49.4k/R_g+1
gain=49.4k/10k+1=5.94
V_{dif} 1.3V-1.2V=0.1V
V_{out}=V_{dif}*gain+V_{ref}
V_{out}=0.1V*5.94+1V=1.59V

Figure 4.10-1

The CMRR (common mode rejection ratio), specifies the level of noise minimization.

One common constraint is the "+" input cannot exceed the "+" power rail by more than 0.5V. Exceeding this limit can result in damage to the instrumentation amp. The outputs may not be linear if the inputs approach either power rail.

The INA826 does not have this restriction. It can handle up to a 40V input beyond the power rails without any damage. The INA826 may not operate properly in this condition, but it will not be damaged.

4.11 Voltage Regulators

Voltage regulators convert a DC voltage into a different DC voltage. They compare the output to an internal reference, and adjust the output to provide a constant output voltage.

The output voltage remains constant, even when the input voltage changes (line regulation).

There are two types of voltage regulators; linear and switching.

Linear regulators operate in a linear area. The output voltage is always lower than the input voltage (Figure 4.11-1).

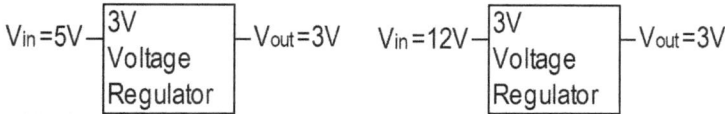

V_{in}=5V ─┤3V Voltage Regulator├─ V_{out}=3V V_{in}=12V ─┤3V Voltage Regulator├─ V_{out}=3V

Figure 4.11-1

The primary advantage of a linear regulator is that the output voltage is "clean" with very little noise.

Some linear regulators allow a very small voltage difference between the input voltage and output voltage. This type of linear regulator is labeled as an LDO (low dropout).

The disadvantage of a linear regulator is power inefficiency. The input current equals the output current regardless of the input voltage and output voltage.

V_{in}=12V, V_{out}=3V, I_{out}=30mA output required
Linear regulator where I_{in}=I_{out}
P_{out}=V_{out}*I_{out}
P_{out}=3V*30mA=0.09w
P_{in}=V_{in}*I_{in}
P_{in}=12V*30mA =0.36w
$P_{efficiency}$=P_{out} / P_{in}
$P_{efficiency}$=0.09w/0.36w=25%

Switching regulators switch on and off. The amount of charge to the load is based on the duty cycle (time on divided by the time on plus time off). A key component in a switching voltage regulator is an inductor. The inductor stores the energy from the on and off switching.

Advantages of switching regulators are the ability to create: 1) an output voltage that is lower than the input voltage, 2) an output voltage that is higher than the input voltage, 3) a negative output voltage, and 4) positive and negative output voltages simultaneously (Figure 4.11-2). The Linear Technology LT1615 generates ±10V from 3V with very few components.

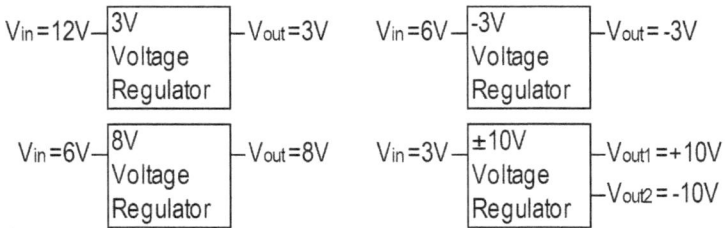

$V_{in}=12V$—	3V Voltage Regulator	—$V_{out}=3V$	$V_{in}=6V$—	-3V Voltage Regulator	—$V_{out}=-3V$
$V_{in}=6V$—	8V Voltage Regulator	—$V_{out}=8V$	$V_{in}=3V$—	±10V Voltage Regulator	—$V_{out1}=+10V$ —$V_{out2}=-10V$

Figure 4.11-2

Another advantage of a switching regulator is power efficiency. If a switching regulator were 100% efficient, the input voltage times the input current would equal the output voltage times the output current (Figure 4.11-3).

$V_{in}=12V$, $V_{out}=3V$, $I_{out}=30mA$ output required
Switching regulator with $P_{efficiency}=100\%$
$P_{out}=V_{out}*I_{out}$
$P_{out}=3V*30mA=0.09w$
$P_{in}=P_{out}/P_{efficiency}$
$P_{in}=0.09w/100\%=0.09w$
$I_{in}=P_{in}/V_{in}$
$I_{in}=0.09w/12V=7.5mA$

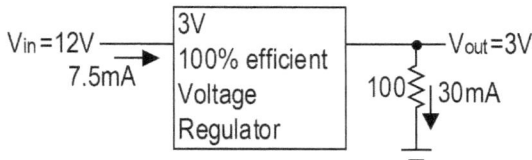

Figure 4.11-3

This can provide a significant effect on the input power requirements, which is often necessary for low power products.

There are always some losses which result in lower power efficiencies (Figure 4.11-4). These losses are often dependent on the output current requirements.

$V_{in}=12V$, $V_{out}=3V$, $I_{out}=30mA$ output required
Switching regulator with $P_{efficiency}=85\%$

$$P_{out} = V_{out} * I_{out}$$
$$P_{out} = 3V * 30mA = 0.09w$$
$$P_{in} = P_{out} / P_{efficiency}$$
$$P_{in} = 0.09w / 85\% = 0.106w$$
$$I_{in} = P_{in} / V_{in}$$
$$I_{in} = 0.106w / 12V = 8.8mA$$

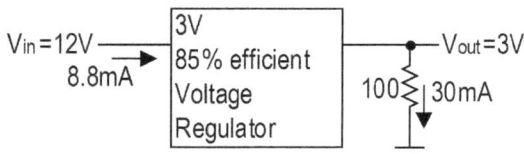

Figure 4.11-4

One disadvantage of switching regulators is noise. The on and off characteristics of switching regulators generates noise. This is rarely a problem for digital electronics. It can be a problem for analog electronics.

PCB layouts are also a disadvantage for switching regulators. The correct location of components is important for the switching regulator to work properly. It is common for switching regulator datasheets to provide layout information.

4.12 Voltage References

Voltage references are used to provide accurate, low noise voltage outputs. This is important for assemblies that use switching voltage regulators (Figure 4.12-1).

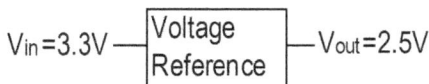

Figure 4.12-1

Voltage references rarely provide more than 1mA of current. They are used to provide an accurate, low noise output voltage that is stable over a wide temperature range.

Some voltage references do provide more current. The REF3025 has an output of 25mA.

4.13 Voltage Supervisors

Voltage supervisors monitor the power supply voltage. When that voltage drops below a specified voltage, the voltage supervisor

generates a pulse that is used to reset micros and other components that are dependent on the power voltage.

The Intersil ISL88002 and the Analog Devices ADM803SAKS
detect when the voltage is below 3.9v. When this occurs, their
output is 0V. This output is connected to a micro reset input. The
0V resets the micro.

Chapter 4 Quiz

4-1   What is the purpose of a comparator?

4-2   It is common to use what with comparators to avoid double
pulses when the input voltage is noisy?

4-3   An op amp "+" input is connected to ground. The "-" input
is connected to a 2.67k resistor to an input voltage. The "-"
input is also connected to the op amp output with a second
resistor. The input voltage is 2V. The output voltage is -3V.
What is the value of the second resistor?

4-4   An op amp "+" input is connected to an input voltage. The
"-" input is connected to a 2k resistor to ground. The "-"
input is also connected to the op amp output with a 4.99k
resistor. The output voltage is 2.6V. What is the input
voltage?

4-5   What is the effect of an op amp input offset voltage?

4-6   What is op amp input bias current?

4-7   What is op amp bandwidth?

4-8   What is op amp slew rate?

4-9   What are the advantages of 0.1% resistors used with
differential input op amps?

4-10   Is it better to provide constant current or constant voltage
to resistive type sensors in series?

4-11   What is CMRR?

4-12   What is a common voltage constraint regarding an instrumentation amp "+" input?

4-13   What type of voltage regulator generates very little noise?

4-14   What are four advantages of switching regulators?

4-15   An 9V input into a switching regulator results in a 3.3V
output. The maximum input current is 3mA. The switch
regulator has an efficiency of 84%. What is the maximum
current allowed out of the switch regulator?

4-16   What is a good reason to use voltage references?

4-17   Is it common for voltage references to provide large amounts of output current?

4-18   What is the primary purpose of a voltage supervisor?

4-19   What is the output voltage of a voltage supervisor when the power supply voltage drops below a specified voltage?

Chapter 4 Design Challenges (use SPICE models when applicable)

4:1   An op amp is powered by 10V. The output is connected to a micro powered by 3V. The maximum input voltage to the micro is 3.5V

Design a circuit that protects the micro using a Zener.

4:2   An analog differential signal is ten feet from the main electronic assembly. The maximum differential signal amplitude is 5mV with a maximum frequency of 10Hz.

Design a circuit with components on the differential signal electronic assembly, and components on the main electronic assembly, that results in 2.5V for an ADC input on the main electronic assembly.

4:3   The input to an electronic assembly will be 0 to 5V, expecting a ≈500 ohm load. An inadvertent connection to this input is 25V.

Design a circuit that protects the electronic assembly using a depletion mode N Channel FET.

Chapter 5

*D*IGITAL

5.1 Digital Component Basics

Digital components reduce potential electronic noise interference. Analog components rely on a wide range of voltages. Electronic noise can produce incorrect results from an analog component.

Digital component I/O (inputs / outputs) is either a "1" or "0". Electronic noise has close to zero effect on the results.

With a power rail of 3.3V (IC is powered by 3.3V) a "1" equals 3.3V and a "0" equals 0V. With a power rail of 5V, a "1" equals 5V and a "0" equals 0V.

For a 3.3V power rail, the "1" does not need to be exactly 3.3V. It can be as low as 2.4V. A "0" is between 0V and 0.8V.

An amplitude between 0.8V and 2.4V produces an uncertain result. It could be interpreted as a "1" or a "0". This should never happen unless there is an error on the PCB.

Some basic digital logic gates (Figure 5.1-1) include; "inverter" (7404), "and" (7408), "nand" (7400), "or" (7432), and "nor" (7402). As an example, an "inverter" inverts the input.

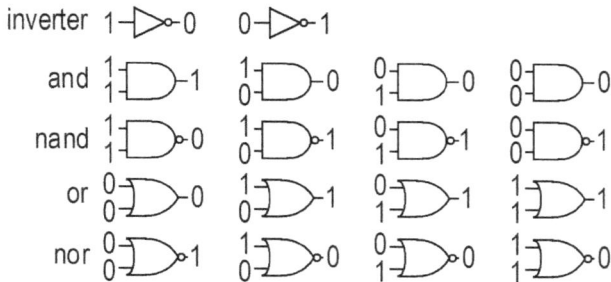

Figure 5.1-1

Open drain logic gate makes it simple to level shift (for example, converting a 3V signal to a 5V signal). Open drain logic gates require pullup resistors (Figure 5.1-2).

Figure 5.1-2

Open collector components also allow the connection of multiple outputs (Figure 5.1-3).

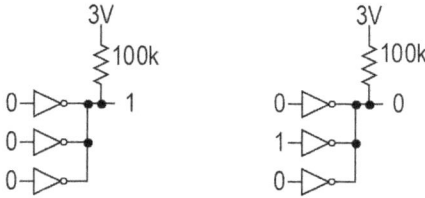

Figure 5.1-3

When mechanical switches change positions, the inputs are momentarily floating. After mechanical switches make contact, the output of the mechanical switches often bounce. Digital logic gates can eliminate this problem (Figure 5.1-4).

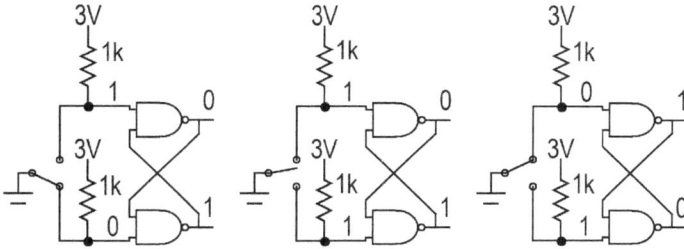

Figure 5.1-4

5.2 Binary Numbers / Hex Numbers

Every binary number is represented by a "1" or a "0". It is common to place the letter "b" at the end of the binary number.

With a two bit binary number, the right digit represents a "0" or "1" times 1. The left digit represents a "0" or "1" times 2.

Two bit binary numbers:

$$00b=(0*2)+(0*1)=0$$
$$01b=(0*2)+(1*1)=1$$
$$10b=(1*2)+(0*1)=2$$
$$11b=(1*2)+(1*1)=3$$

Four bit binary numbers:

$$0000b=0+0+0+0=0 \qquad 1000b=8+0+0+0=8$$
$$0001b=0+0+0+1=1 \qquad 1000b=8+0+0+1=9$$
$$0010b=0+0+2+0=2 \qquad 1000b=8+0+2+0=10$$
$$0011b=0+0+2+1=3 \qquad 1000b=8+0+2+1=11$$
$$0100b=0+4+0+0=4 \qquad 1000b=8+4+0+0=12$$

0101b=0+4+0+1=5   1000b=8+4+0+1=13
0110b=0+4+2+0=6   1000b=8+4+2+0=14
0111b=0+4+2+1=7   1000b=8+4+2+1=15

The standard is to use eight bit binary numbers (bytes):

00000000b=0   00010000b=16
00000001b=1   00100000b=32
00000010b=2   01000000b=64
00000100b=4   10000000b=128
00001000b=8   11111111b=255

As it can be difficult to interpret a large group of "1" and "0" bits, a hex nomenclature is used. It is common for hex number to be preceded by "0x". The numbers 10 through 15 are replaced by letters of the alphabet:

0000b=0=0x0   1000b=8=0x8
0001b=1=0x1   1001b=9=0x9
0010b=2=0x2   1010b=10=0xA
0011b=3=0x3   1011b=11=0xB
0100b=4=0x4   1100b=12=0xC
0101b=5=0x5   1101b=13=0xD
0110b=6=0x6   1110b=14=0xE
0110b=7=0x7   1111b=15=0xF

00011010b=0x1A
11001000b=0xC8
11111111b=0xFF

5.3 Bytes, Integers, Long, Unsigned

Bits are often in groups of 8, 16, or 32. The MSB (most significant bit) is on the left. The LSB (least significant bit) is on the right.

unsigned: all bits are a positive number
signed: the MSB is a sign bit, the remainder are a number

unsigned byte (also known as a char): 8 bits
11111111b=0xFF=255
00000101b=0x05=5

signed byte 8 bits
11111110b=0xFE=254

00000101b=0x05=5

unsigned integer (unsigned int): 16 bits
1111111111111111b=0xFFFF=65535
0100001010111000b=0x42B8=17080

signed integer (int): 1 sign bit (MSB) and 15 data bits
0111111111111111b=0x7FFF=32767
1000000000000000b=0x8000= -32768
1111111111111101b=0xFFFD= -3
1111111111111110b=0xFFFE= -2
1111111111111111b=0xFFFF= -1
0000000000000000b=0x0000=0
0000000000000001b=0x0001=1

unsigned long (unsigned long): 32 bits
11111111111111111111111111111111b=0xFFFFFFFF=
4294967295
01110100111011001000101110100001b=0x74EC8BA1=
1961659297

signed long (signed long): 1 sign bit (MSB) and 31 data bits
01111111111111111111111111111111b=0x7FFFFFFF=
2147483647
10000000000000000000000000000000b=0x80000000=
-2147483648
11111111111111111111111111111111b=0xFFFFFFFF=-1
00000000000000000000000000000001b=0x00000001=1

Note that there some variations where "short" is 16 bits, "int" is 32 bits, and "long" is 64 bits.

5.4 Float, Double

Sometimes there is a need for higher precision which includes decimal points (floating point). A "float" provides precision. A "double" doubles that precision.

IEEE 754 Floating Point Standard
float: 32 bits
1 bit sign (s) 8 bits exponent (e) 23 bits mantissa (m)
seeeeeeeemmmmmmmmmmmmmmmmmmmmmmm

double: 64 bits
1 bit sign (s) 11 bits exponent (e) 52 bits mantissa (m)
seeeeeeeeeeemmmmmmmmmm…mmmmmmmmmmmm

decimal	float: 32 bits	double: 64 bits
0	0x00000000	0x0000000000000000
1	0x3F800000	0x3FF0000000000000
1.1	0x3F8CCCCD	0x3FF199999999999A
1.2	0X3F99999A	0x3FF3333333333333
2	0X40000000	0x4000000000000000
100	0X42C80000	0x4059000000000000
-1	0XBF800000	0xBFF0000000000000

When using micros, avoiding floating point numbers consumes less memory space and takes less computational time.

The Internet provides detailed descriptions of the computations to create floating point numbers.

5.5 ADCs (Analog to Digital Converters)

ADCs convert analog voltages into binary numbers.
Two bit ADC:

Range: from 0V to 5V	*Range: from 0V to 13V*
00b=5V/3*0=0V	*00b=13V/3*0=0V*
01b=5V/3*1=1.667V	*01b=13V/3*1=4.333V*
10b=5V/3*2=3.333V	*10b=13V/3*2=8.667V*
11b=5V/3*3=5V	*11b=13V/3*3=13V*

Four bit ADC:

Range: from 0V to 3V	
0000b=3V/15*0=0.0V	*1000b=3V/15*8=1.6V*
0001b=3V/15*1=0.2V	*1001b=3V/15*9=1.8V*
0010b=3V/15*2=0.4V	*1010b=3V/15*10=2.0V*
0011b=3V/15*3=0.6V	*1011b=3V/15*11=2.2V*
0100b=3V/15*4=0.8V	*1100b=3V/15*12=2.4V*
0101b=3V/15*5=1.0V	*1101b=3V/15*13=2.6V*
0110b=3V/15*6=1.2V	*1110b=3V/15*14=2.8V*
0111b=3V/15*7=1.4V	*1111b=3V/15*15=3.0V*

The conversion from input voltages to output binary numbers requires an input reference voltage. The input voltage when compared to the reference voltage produces an output binary number (Figure 5.5-1).

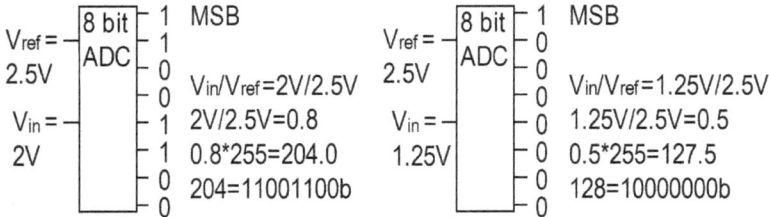

Figure 5.5-1

5.6 DACs (Digital to Analog Converters)

DACs convert input binary numbers into output voltages. This conversion requires an input reference voltage (Figure 5.6-1).

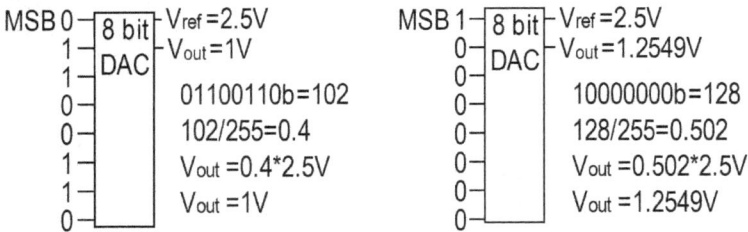

Figure 5.6-1

5.7 Micros (Microcontrollers / Microprocessors)

Micros are components that can be programmed to perform specific functions. **Code** is the binary numbers that provide those instructions to the micro.

Using binary numbers as an input, micros interpret those numbers and take action. Microprocessors have more capability than microcontrollers.

The code for a simple microcontroller is labeled as firmware. Code for a more complex microprocessor, as is found in computers, is labeled as software.

A common term for microcontrollers is the word embedded.

5.8 Watchdog Timers

Watchdog timers detect when micros cease to function properly.

During normal operation, a micro issues periodic pulses to the watchdog timer. When the pulses stop, the watchdog timer resets the micro.

5.9 Memory

There are two types of memory used by micros, volatile and non-volatile.

RAM is volatile memory. RAM is used by micros as a temporary storage place. When the electronic assembly is powered down, the contents of the RAM are erased.

FLASH is NVM (non-volatile memory). When the electronic assembly is powered down, the contents of the FLASH are not erased.

FLASH contains firmware code and information that needs to be preserved when power is removed. Micros rely on FLASH for code instructions. Code dictates the actions of micros.

Other types of NVM include EEPROM, magnetic tape, floppy disks, and hard disks.

The primary limitation of FLASH / EEPROM is that a segment (multiple storage locations) must be erased when a change is made in the data. It is not possible to erase, or rewrite, only one byte.

RAM does not have this limitation. Individual bytes can be rewritten without affecting other bytes. RAM is faster than FLASH / EEPROM.

FRAM (or MRAM) is an NVM exception. It is NVM, but can be accessed as RAM is accessed. When FRAM is powered down, the contents are not erased.

There is a limit to the number of times FLASH / EEPROM can be rewritten. It is approximately 24,000 times.

There is no limit to the number of times RAM can be rewritten. While there is a limit to the number of times FRAM can be rewritten, the number for FRAM is in the millions.

The primary disadvantage of FRAM is that it takes more real estate than FLASH / EEPROM and is more expensive. When space is a key limitation, it usually works best to have FLASH /

EEPROM for the bulk of the code, and FRAM for the storage of parameters that may change when the assembly is operational.

Micron / Numonyx M25P16 and Winbond W25Q16 are FLASH. Ramtron / Cypress FM25CL64B is FRAM.

5.10 Firmware / Software Code

Code is written in a readable syntax. The readable syntax is converted to binary numbers using a compiler. Micros require binary numbers.

A common programming language for micros is C. A compiler converts C code into Assembler code, which is then converted into binary numbers. Binary numbers are stored in FLASH. Those binary numbers dictate the actions of the micro.

Texas Instruments MSP430 code

C code   *uint8_t ucValue=1;*
Assembler code   *mov.b #0x1, R10*
Flash adr   *0x6DC2 (instruction at adr 0x6DC2 is 0x435A)*

C code   *uint16_t uiValue=2;*
Assembler code   *mov.w #0x2, R11*
Flash adr   *0x6DC4 (instruction at adr 0x6DC4 is 0x432B)*

C code   *uint16_t uiSum=ucValue+uiValue;*
Assembler code   *mov.b #R10,R8*
Flash adr   *0x6DC6 (instruction at adr 0x6DC6 is 0x4A48)*
Assembler code   *add.w #R11,R8*
Flash adr   *0x6DC8 (instruction at adr 0x6DC8 is 0x5B08)*

The micro has a PC (program counter) that dictates the next instruction to execute. In the code example, the PC changes from 0x6DC2 to 0x6DC4 after executing the instruction at 0x6DC2.

Some code writers use title case for variables. Title case is comprised of concatenating words. It is easier to read "Position-GoalPercent" than "POSITIONGOALPERCENT".

Some code writers use Hungarian notation for variables. Hungarian notation adds characters to the front of the variable name.

The variable name "dGoal" can mean that variable is defined as a "double". The variable name "aucPulseCount" can mean that variable is defined as an "array of unsigned chars".

There is no standard for Hungarian notation.

5.11 Parallel Communication

Parallel communication is access to multiple bits simultaneously.

In Figure 5.11-1, the 8 bits (byte) of data change on the rising edge of Clock. The data does not change (is valid) on the falling edge of Clock.

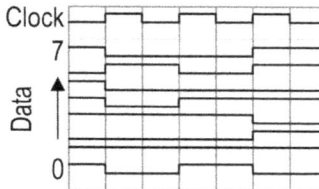

Figure 5.11-1

The first byte is 01001010b (0x4A), second byte is 00011011b (0x1B), third byte is 11010110b (0xD6).

5.12 SPI Serial Communication

Serial communication is access to one bit at a time.

SPI (Serial Peripheral Interface) is a standard serial communication protocol. It is comprised of one master and one, or more, slaves. The data is an MSB format.

A master outputs SpiClk (clock), SpiMoSi (master out serial in), and CS (chip select). A slave outputs SpiMiSo (master in slave out).

CS is used to enable slave components. When there are multiple slaves, there is one CS for each slave. A slave is activated when their CS is "0". Slaves output SpiMiSo only when their CS is selected.

In Figure 5.12-1, SpiMoSi changes on the falling edge of SpiClk. This means the data is valid on the rising edge of SpiClk. CS is "0" before SpiClk starts and "1" after SpiClk stops.

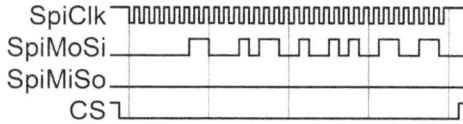

Figure 5.12-1

The first byte on SpiMoSi is an op code of 0x03 (indicates to the slave this is a write command). The second byte is an address of 0x16 (indicates to the slave the location to place the first data byte). The third and fourth byes are data bytes 0x4A and 0x66.

In Figure 5.12-2, the first SpiMoSi byte is an op code of 0x05 (indicates to the slave this is a read command). The second SpiMoSi byte is an address of 0x31. The third and fourth byes are data bytes 0x8E and 0x47 from the slave on SpiMiSo.

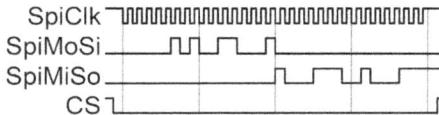

Figure 5.12-2

When additional SpiClks are sent, the slave continues to place bytes on SpiMiSo. In Figure 5.12-2, the second and third data bytes would be in a sequential order continuing with addresses 0x32 and 0x33.

5.13 I2C Serial Communication

I2C is a serial communication that requires only two connections, SCL (serial clock) and SDA (serial data). They are driven with open collector drivers and pull-up resistors.

Comprised of one master and one or more slaves, I2C data is in an MSB format.

SCL is controlled by the master. SDA starts with master control. SDA is controlled by a slave at the appropriate time.

The master starts the process by issuing a Start. The master also issues the Stop. The Start and Stop occur by changing SDA while SCL is "1". All other SDA changes occur while SCL is "0" (Figure 5.13-1).

There is no CS for I2C. A slave address identifies which slave should interpret the data. After the master sends a slave address and a read / write bit, that slave responds with a 1 bit acknowlededge.

For a write, the slave responds with a 1 bit acknowledge for each byte of data sent by the master.

In Figure 5.13-1, the slave address is 0x4E, the data address is 0x6A, and the data is 0x74. There could be additional bytes sent after the 0x74 by continuing SCL before issuing a Stop bit.

Master writing to a Slave
Sadr: 7 bit slave address sent by Master
Dadr: 8 bit data address sent by Master
Data: 8 bit data sent by Master
Ack: 1 bit acknowledge sent by Slave for each byte

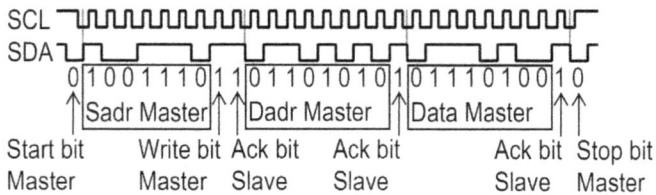

SCL
SDA
0 1 0 0 1 1 1 0 1 1 0 1 1 0 1 0 1 0 1 0 1 1 1 0 1 0 0 1 0

↑Sadr Master↑ ↑Dadr Master↑ ↑Data Master↑↑

Start bit   Write bit Ack bit   Ack bit   Ack bit Stop bit
Master   Master   Slave   Slave   Slave   Master

Figure 5.13-1

For a read, the master responds with a 1 bit acknowledge for each byte of data sent by the slave.

In Figure 5.13-2, the slave address is 0x2D, the data address is 0x4B, and the data is 0x16. There could be additional bytes sent after the 0x16.

Master reading from a Slave
Sadr: 7 bit slave address sent by Master
Dadr: 8 bit data address sent by Master
Data: 8 bit data sent by Slave
Ack: 1 bit acknowledge sent by Slave or Master

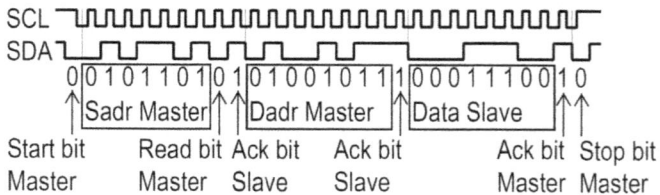

SCL
SDA
0 0 1 0 1 1 0 1 0 1 0 1 0 0 1 0 1 1 1 0 0 0 1 1 1 0 0 1 0

↑Sadr Master↑ ↑Dadr Master↑ ↑Data Slave↑↑

Start bit   Read bit   Ack bit   Ack bit   Ack bit Stop bit
Master   Master   Slave   Slave   Master   Master

Figure 5.13-2

The details about the protocol may vary between different ICs. Datasheets define the protocol for that component.

5.14 RS-232 / RS-485 Serial Communication

Serial communication can also be accomplished with asynchronous communication. This means there is no master clock.

One standard asynchronous communication protocol is RS-232. This was an interface to personal computers for many years (Figure 5.14-1).

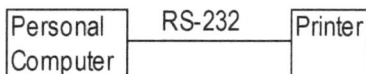

Figure 5.14-1

Standard RS-232 frequencies are 2400, 4800, 9600, 19200, and 38400. For a frequency of 9600 bits per second, each bit is spaced by 104.1667 microseconds (1/9600).

RS-232 requires a Start bit and a Stop bit. The Start bit is a positive voltage, and Stop bit is a lower voltage. The data bits are in groups of 8 bits (some protocols use fewer than 8 bits).

The "1" bits and "0" bits are inverted, where "0" bits are a higher voltage than "1" bits. The LSB is sent first.

In Figure 5.14-2, the bytes are 0x03 (00000011b), 0xFE (11111110b), and 0x64 (01100100b).

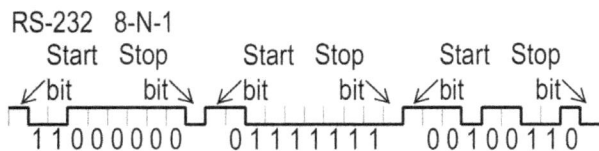

Figure 5.14-2

A parity bit can be required. Parity bit types are odd, even, or none. When the parity bit type is odd or even, the parity bit is added after the data bits.

The letters 8N1 indicate 8 data bits, no parity bit, and 1 stop bit. The letters 8O1 indicate 8 data bits, odd parity bit, and 1 stop bit.

For odd parity, the number of "1" bits, including the parity bit must be an odd number of "1" bits. In Figure 5.14-3 the parity bit is "1" for 0x03, and the parity bit is "0" for 0x64.

RS-232 8-O-1

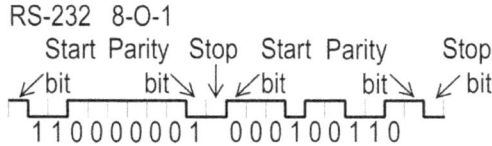

Figure 5.14-3

The RS-232 transmitter and receiver must be set to the same frequency, the same number of data bits, and the same parity type.

RS-232 originally started with a 25 pin connector. Later, a 9 pin connector became standard. The 9 pin connector was used consistently for personal computers.

RS-232 9 Pin Connector
Pin 1: Data Carrier Detect   *Pin 6: Data Set Ready*
Pin 2: Received Data   *Pin 7: Request to Send*
Pin 3: Transmit Data   *Pin 8: Clear to Send*
Pin 4: Data Terminal Ready   *Pin 9: Ring Indicator*
Pin 5: Ground

With improved technology, only three pins are used; Receive Data, Transmit Data, and Ground.

A variation of RS-232 is to use duplex or half-duplex. Full-duplex uses one wire for transmit and another wire for receive. Half-duplex uses the same wire for transmit and receive.

As this is digital data, the Receive Data and Transmit Data are either a "1" or a "0". Over long distances, noise can affect the performance. To help reduce this problem, the RS-422 (differential communication) became a standard for communication.

For a "1", one wire is a higher voltage than the other wire. For a "0" the other wire is higher voltage. Comparing one wire to the second wire determines if it is a "1" or a "0".

RS-485 is an extension of RS-232. An RS-485 allows connecting multiple devices to the same pair of wires. There is an address with RS485 communication.

Communicating over a long length of wire with RS-232 and RS-485 can cause voltage reflections. The reflection amplitude can be reduced by using a terminating resistor.

The Maxim MAX3483E provides a convenient way to communicate RS-485.

As RS-232 connectors on computers are becoming less common, there are devices that convert USB to RS-485.

5.15 Little Endian / Big Endian

When groups of bytes (int, long, float, double) are sent serially, are they sent little endian or big endian?

Assume the number is 0x1A. The 0x1 byte is sent first for big endian. The 0xA byte is sent first for little endian. Another consideration is the bit order (MSB or LSB).

For RS-232, the LSB is sent first. For SPI and I2C, the MSB is sent first.

Chapter 5 Quiz

5-1   What is the primary purpose of digital logic?
5-2   What does level shift mean?
5-3   What is the hex number for 01111010b?
5-4   What is the decimal number for 0xA8?
5-5   What is the advantage of an unsigned number?
5-6   What is the disadvantage of an unsigned number?
5-7   What is the advantage of a floating point number vs. an integer number?
5-8   What is the advantage of a double number vs. a float number?
5-9   What is the purpose of an ADC?
5-10   An 8 bit ADC has a voltage reference of 3V and an output of 0xFE. What is the input voltage?
5-11   What is the purpose of a DAC?
5-12   A 12 bit DAC, as a voltage reference of 3.3V and an input of 0xBB. What is the output voltage?
5-13   What is the primary advantage of FLASH vs. RAM?
5-14   What are two primary advantages of RAM vs. FLASH?
5-15   What is the advantage of FRAM vs. RAM?
5-16   What are two disadvantages to of FRAM vs. FLASH?
5-17   What is the purpose of a compiler?
5-18   What is the function of a PC?
5-19   What is the advantage of parallel communication?
5-20   What is the disadvantage of parallel communication?
5-21   What is the purpose of CS for SPI?
5-22   How many connections are required for SPI read / write communication to multiple devices?

5-23   How many connections are required for I2C?

5-24   Why does an I2C slave not require a CS?

5-25   What does asynchronous mean?

5-26   What is the purpose of a parity bit?

5-27   What is the advantage of RS-422 vs. RS-232?

5-28   What is the difference between RS-422 vs. RS-485?

5-29   Given 0xFE. Which byte is sent first for little endian?

Chapter 5 Design Challenges (use SPICE models when applicable)

5:1   A double throw switch controls an electronic assembly. The pullup resistors to the switch contacts are connected to 3V.

   Design a circuit that provides an appropriate interface to a micro powered by 3V using digital logic.

5:2   A double throw switch controls an electronic assembly. The pullup resistors to the switch contacts are connected to 3V.

   Design a circuit that provides an appropriate interface to a micro powered by 3V using digital logic that does not use nand gates.

Chapter 6

EMC

6.1 EMC Design Considerations

EMC (electro-magnetic compatibility) is a branch of electronics about unintentional generation, propagation, and reception of electro-magnetic energy. EMC requires appropriate design of the electronic assembly and may require appropriate design of the cables and the enclosure.

Metallic shielding of the electronic assembly and shielding and / or filtering of the cables may be necessary to meet EMC requirements. This approach is often used in harsh environments (military), and in industrial applications.

When this adds significant cost to the product, additional EMC control at the electronic assembly may be the only option.

Important elements in EMC control at the assembly level are minimizing ground impedance, decoupling, limiting rise / fall times of pulses, and I/O filtering.

The basic need for ground is low impedance from one point to any other point. Ground planes should have exceedingly low impedances and are preferred to meet EMC requirements.

Wires and PCB traces are inductive. This can lead to high impedances at high frequencies.

The inductance in a wire is approximately 20nH/inch (8nH/cm). A one inch length of wire may have an inductance of 20nH, and an impedance of approximately 12Ω at 100MHz ($Z=2*\pi*$frequency$*$inductance). An impedance of 12Ω is usually too large.

Ground impedance needs may vary, but generally impedances above 1Ω are too large. Impedances of ground planes are usually in the low milliohm range.

High frequency shielding is mandatory for military and avionics equipment. High frequency shielding is often implemented for other equipment as it simplifies the electronic design.

The most stringent aspect of shielding is usually not the thickness and conductivity of the shield material. The most stringent aspect is the openings and cable penetrations in the electronic assembly housing.

The maximum allowed opening is 1/20 the wavelength of the highest frequency of interest. Wavelength is defined by lambda, where lambda (meters) equals 300/frequency in MHz.

For emissions, the highest frequency of interest (fm) is ten times the highest clock frequency. For immunity, fm is usually 1GHz or higher. For surge transient, fm is 300kHz. For electrical fast transient, fm is 60MHz. For electrostatic discharge fm is 300MHz.

It is common for power cables and data cables to be filtered or shielded. For filters, $0.01\mu F$ capacitors connected to earth ground are usually sufficient.

Cable shielding is primarily driven by the shield termination, and conductive termination on all mating surfaces from the cable shield to the connector back shell.

Pigtail connections are usually unacceptable. High frequency cable shielding should be grounded, at both ends, to earth ground. A single point ground is not recommended above audio frequencies.

Ground loops can occur when grounds are connected in two remote locations. A ground loop is a low frequency problem, where the length of the path in question is less than a quarter of a wavelength of the highest frequency of interest.

Generally, the lambda / 20 rule is used for the boundary between low frequencies and high frequencies. This is also approximately the boundary between lumped and distributed circuit behavior.

The key assumption in single point grounds is that the speed of light is infinite, so that events happen simultaneously in all components of a system. Once the time delay from one corner to another is considered, it can no longer be assumed that a single point ground will work.

When working with audio frequencies, including 60Hz, single point grounds are appropriate. However, when the path to ground is a 1/4 wavelength, then a ground may not exist.

In addition, at high frequencies, stray capacitive paths defeat the single point ground. At radio frequencies, it is best to ground both ends (well known by radio designers).

When dealing with both high and low frequencies, it works well to have a hybrid ground (hard ground on one end and a capacitive ground at the other end).

It also works well to have double shielding (outer shield grounded at both ends, inner shield grounded at one end). The outer shield intercepts the high frequencies; the inner shield intercepts the low frequencies created by the ground loop.

Crucial issues in the electronic assembly design, at the PCB level, are solid ground planes with power decoupling and filtering of I/O connections.

Decoupling capacitors ($0.01\mu F$ to $0.1\mu F$) are important at all power connections and at the PCB boundary. Decoupling capacitors are important for all I/O connections. The distance of the decoupling capacitors between earth ground and connections should be kept to an absolute minimum length.

Placement of electronic components on the PCB is also important. It is best to keep pulses with fast rise / fall times away from the I/O connectors.

For regions where ground planes are not an option, it works best to keep trace lengths as short and fat as possible, and by interconnecting the ground traces wherever possible.

6.2 TVSs (Transient Voltage Suppressors)

Fast voltage spikes can damage electronic components. One voltage spike source is lightning.

The voltage spikes are rarely more than a few microseconds, but can be a few thousand volts.

Solutions include a TVS which may be a large package Zener diode, an MOV (metal-oxide varister), or an arc suppression component. These components respond quickly, to limit voltage magnitudes.

A MSMBJ12CA is a TVS. It provides some protection for voltages that are designed to use 12V. The MSMBJ12CA starts to break down between 13.3V and 14.4V. It will start providing a path to ground between these voltages.

Because a TVS cannot respond instantly, the voltage can reach a higher voltage for a short period of time. The MSMBJ12CA maximum clamping voltage is 19.9V. This means a voltage spike as high as 19.9V could occur for less than a microsecond, and then drop to the break down voltage.

For many I/O connections the maximum voltage spike applied for testing is 1,000V. An MSMBJ12CA, rated for 600w, can usually handle this.

For power rails connections, the maximum voltage spike applied for testing could be 2,000V. An SMT15T12CA can usually handle this as it is rated for 1500w.

The "CA" at the end of the component number indicates it is bidirectional. It can handle positive and negative voltage spikes. Components with "A" at the end of the component number indicates it can handle only positive voltage spikes or negative voltage spikes, not both.

6.3 Common Mode Chokes

Common mode chokes can be effective for EMC. The section on Common Mode Chokes covers these details.

6.4 Ferrite Cores

Ferrite cores are composed of a high permeability, ferromagnetic material surrounding a wire. They are a high frequency lossy inductor, effectively restricting high frequency currents from traveling through the wire by slowing the rise time of current pulses.

The composition and size affect different frequencies. The most common ferrite is a nickel zinc formulation, providing effective interference control in the 30 to 1GHz range, which equates to approximately 1 nanosecond rise time.

Ferrite cores are often used in electric motors to reduce EMC emissions, commonly placed on the wires to the motor. They are most effective when placed at noise sources to prevent wires from being antennas.

Chapter 6 Quiz

6-1   What is EMC?
6-2   Are EMC requirements about the radiation of equipment or susceptibility of equipment?
6-3   What is the purpose of a TVS?
6-4   A TVS is similar to what component?
6-5   What is the primary characteristic of a common mode choke?
6-6   What is the primary advantage of a common mode choke?
6-7   What effect does a ferrite core have on current?
6-8   A ferrite code is similar to what electronic component?

Chapter 7

ELECTRONICS IMPORTANT DETAILS

7.1 Datasheets

There are multiple aspects to electronic design. These details include effect of temperature, parameter tolerances, and worst case power (which can cause the device to overheat).

A component manufacturing company knows what parameter changes could occur in the components they sell. These parameter changes could be different from lot to lot (components manufactured one month could be slightly different from components manufactured the next month).

Datasheets are crucial to electronic design. They define parameter changes that could occur. Bench-top testing can confirm circuit designs, but should not dictate the design.

While bench-top testing can indicate actual performance for that lot of components, future bench-top testing of another lot of components could provide slightly different results.

7.2 SPICE Models

SPICE model programs (Micro-Cap, LTspice, and PSPICE) can provide important information that fairly accurately predicts end results.

One advantage of a SPICE model is that it can estimate worst case results by varying all component tolerances and by producing anticipated results as a function of temperature.

It could be difficult and time consuming to change every component with bench top testing. SPICE models provide insight into what may happen.

Information about free SPICE programs and SPICE models can be found at http://ElectronicsIsEasy.com. Appendix A and Appendix B provide additional information about Micro-Cap and LTspice.

Example 1: Temperature (Figure 7.2-1):

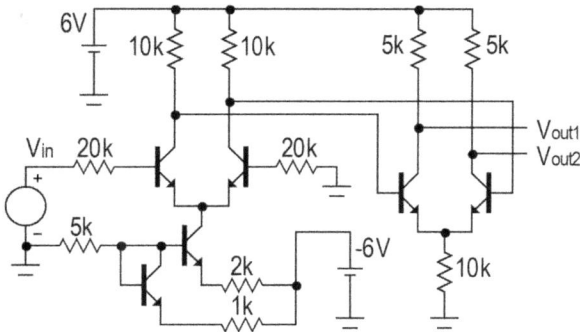

Figure 7.2-1

The output voltage (V_{out1}-V_{out2}) changes with temperature. The larger voltage, in Figure 7.2-2, is the output at 125 °C. The middle voltage is the output at 27 °C. The smaller voltage is the output at -55 °C.

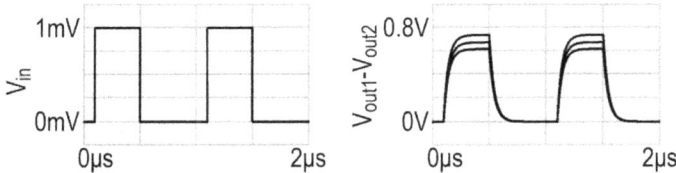

Figure 7.2-2

Example 2: Stepping Component Values (Figure 7.2-3):

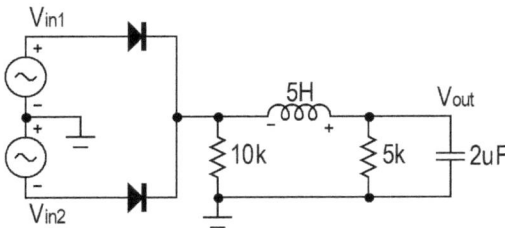

Figure 7.2-3

The inputs are opposite voltages. The three outputs represent the three different resistor values (Figure 7.2-4).

The 5k resistor value is stepped from 5k to 25k in 10k increments. The V_{out} top trace, in Figure 7.2-4, is the 25k value, the middle trace is the 15k value, and the bottom trace is the original 5k value.

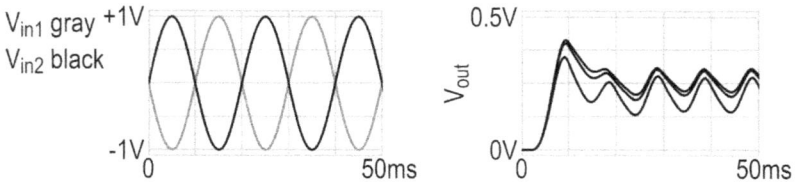

Figure 7.2-4

Example 3: Optimizing a Component (Figure 7.2-5).

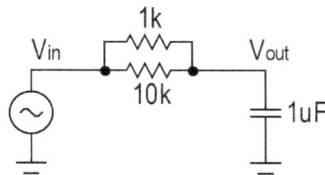

Figure 7.2-5

The optimizer calculates the optimum value for the 1k resistor so the 3dB occurs at 100Hz. The optimizer calculates the value to be 1.887461k.

Example 4: Active Filter Design (Figure 7.2-6).

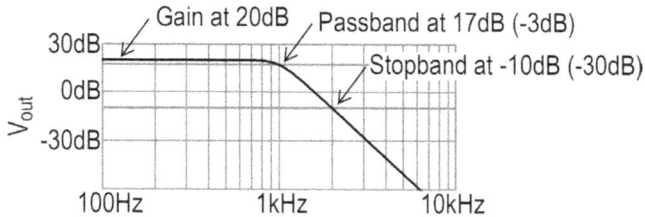

Figure 7.2-6

Design a low pass filter with 20dB Gain, 30dB Stopband Attenuation, 1kHz Passband, and 2kHz Stopband.

Selecting a Butterworth filter, Micro-Cap generates the components for a Butterworth filter (Figure 7.2-7). The component values are not standard values. They can be set to standard values and with Micro-Cap produce the filter output with those changes.

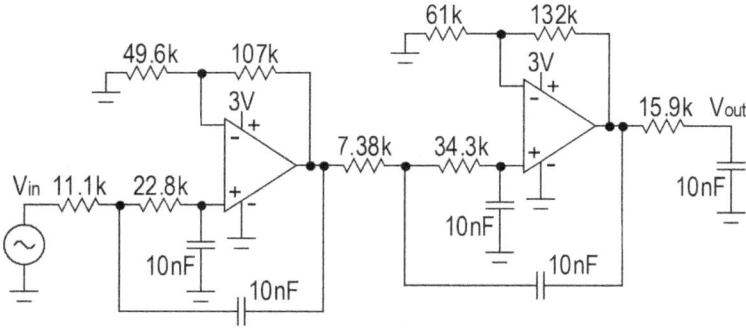

Figure 7.2-7

Example 5: Monte Carlo / Tolerances (Figure 7.2-8).

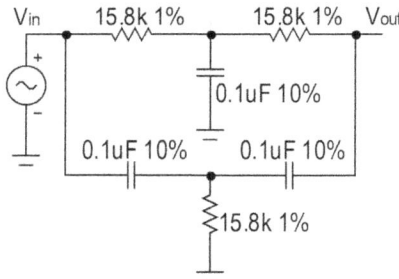

Figure 7.2-8

Determine the effect of component tolerance on a Twin-T filter. Monte Carlo methods are algorithms that use repeated random samples to obtain results.

Figure 7.2-9 uses a Worst Case distribution with 10 samples. It could have used a Uniform or Gauss distribution.

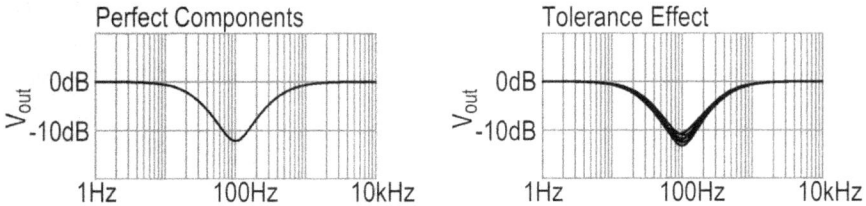

Figure 7.2-9

7.3 Counterfeit Components

Another important detail about electronic design is counterfeiting. Because of the large market of electronic components, some people will sell counterfeit components that may perform similarly to a legitimate component, but may fail over temperature or time.

It is important to establish a "trail" that verifies the components used in manufacturing are not counterfeit components. The best practice is to purchase from factory authorized sources, not from the open market.

Most SMT components are mounted on a tape which is wound around a reel. Some are in tubes. In both cases, there are labels that provide information about the company, part number, manufacturing lot number, and date.

When these components are used, this information is scanned into a computer, along with information about assemblies on which they were mounted. This type of logging system allows tracking of every assembly, and every component on that assembly.

7.4 PCBs (Printed Circuit Boards)

PCBs are used for mounting electronic components. They are comprised of fiberglass with copper foil on the fiberglass. FR4 is a common board substrate. The copper foil is a metal trace that acts as a wire and provides a way for electronic current to flow from one electronic component to another electronic component.

Multiple metal traces allow for interconnecting multiple components. Because the metal traces sometimes overlap, a plated hole (via) is used to provide an electric path from one side of the PCB to the other side or to a layer within the PCB.

A via is the metal tube that connects different layers. These vias are created using an electroplating process. This provides a conductive path between layers.

Metal foils have pads (a metal foil section that is large enough for mounting a component), or metal plated holes. The pads and holes vary in size. They also accommodate the component size and component pin locations.

Metal plated holes are used for mounting PTH components. Pads are used for mounting SMT components.

PCBs can be one layer (metal traces on one side), two layer (metal traces on top side and bottom side), or multi-layer (metal traces on top side, bottom side, and in-between layers).

Many PCBs are in a multi-layer configuration. This is usually due to necessity. It is best to use one inner layer as a voltage plane (most of the layer is a solid voltage plane) and another inner layer as a ground plane (most of the layer is a solid ground plane).

The top and bottom layers are comprised of components and connections between the components. Often two or more inner layers are comprised of connections between components.

Eight layer and ten layer PCBs are common. The density of the components and the number of connections dictate the number of layers.

A solder mask, placed on the PCB, covers the entire PCB except for places that will be soldered. This keeps the PCB copper from oxidizing.

For decades, a thin layer of solder was placed on the exposed copper pads, areas not covered by the solder mask. The entire PCB was passed between a hot air knife (a constant flow of hot air across the PCB surface).

This is called HASL (hot air solder leveling). HASL is used to make the solder on the pads smooth and level.

As SMT components became smaller, and ICs had smaller pitch pins (spacing between pins), HASL surfaces were no longer level enough. To have smoother pad surfaces, electroless nickel immersion gold is used. Electroless nickel immersion gold is a more expensive process but provides a very smooth surface.

Other PCB finishes include OSP (organic surface protection) and immersion (ENIG) silver provide smooth surfaces. These finishes may cost less but have field use concerns.

HASL technology has improved to where the surface is much smoother and works much better for fine pitch components.

Silk screen, which places words or numbers / letters on the solder mask, is often used for prototypes. When products are released for production, silk screen may not be applied to reduce costs.

7.5 Ground Loops

Ground loops can occur when the grounds are connected at two locations. Placing ground connections at only one location eliminates ground loops (Figure 7.5-1).

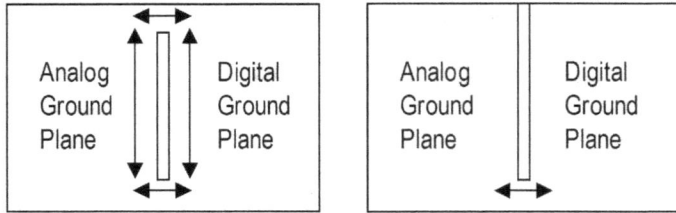

Figure 7.5-1

An exception is ground planes (and voltage planes). A ground plane is a solid section of PCB metal with holes for vias and PTH components. There are no known ground loops for voltage planes and ground planes.

7.6 PCB Fabrication

The precise location of the component pads, vias, test pads, fiducials, and holes on a PCB is crucial.

A PCB must have minimal flexibility. If too flexible, the components can crack or the solder joints can crack when the PCB is flexed.

Locations near assembly mounting screw holes are vulnerable to PCB flexing. It is good to use components that have flexible connections for large size components. The AVX Flexiterm Series capacitors and the Kemet KPS Series capacitors have this flexibility.

Each layer of a multilayer PCBs must be clean. If not clean, there can be dendrite growth between the metal traces.

Dendrite growth is usually not discovered until the assembly has been used by customers for a few months. It is good to qualify a PCB vendor prior to using them as a supplier for products.

Cleaning is also important for external layers before a solder mask is placed on the PCB.

7.7 Solder

Solder is a unique metal alloy. Most solders melt at temperatures below 200 °C. Because of this characteristic, electronic components are attached to PCBs with solder.

Solder is conductive with a very low resistance. This capability allows electronic current to flow between electronic components.

For many decades, solder was comprised of a combination of lead and tin. One of the most common was a ratio of 60% tin and 40% lead. This melts at approximately 183 °C. This makes it relatively easy to connect an electronic component to a PCB.

In 2006, new European regulations banned the use of lead in electronics. This led to the development of solder that uses no lead (mostly tin or tin alloys). This led to an increase of soldering temperatures of 5 to 20 °C.

Solder flux is often mixed with the solder. Flux is a chemical cleaning agent that usually removes contaminants to ensure a good solder connection between the component and the PCB pads.

It is important to remove this flux to prevent contamination which could accumulate over time. This contamination could create an unwanted resistive path between electronic components.

Solder fluxes are currently available in three options:

- *Water soluble flux needs to be removed with simple fluids such as water.*
- *Low solids flux (no-clean process) is a flux that is non-conductive and non-corrosive and may not need to be removed with water. However, it can leave a residue if not fully activated by being heated to a temperature in excess of 100 °C. This can create leakage paths on the PCB.*
- *Rosin flux is a corrosive flux used primarily for existing oxides on a PCB. Cleaning after soldering is preferred.*

7.8 Tin Whiskers

The European Union banned the use of lead in electronic products in the early 21st century. The rest of the world accepted this requirement.

One fact, well known by much of the electronic community, is that the removal of lead from solder and from electronic components can result in tin whiskers.

Tin whiskers are small metal hairs that grow from tin surfaces over time. These metal hairs can create inadvertent, unwanted connections between electronic components. This can cause electronic products to fail.

A document, compiled by Northrop Grumman in 2004, lists the components that use lead.

80.81%	*Storage batteries (car batteries)*
4.78%	*Paints, ceramics, pigments, chemicals*
4.69%	*Ammunition*
1.79%	*Sheet lead*
1.40%	*Cable covering*
1.13%	*Casting metals*
0.72%	*Brass / bronze billets and ingots*
0.72%	*Pipes / traps, extruded products*
0.70%	*Solder (excluding electronic solder)*
0.49%	*Electronic solder*
2.77%	*Miscellaneous*

A small amount of lead, approximately 3%, would eliminate the tin whisker problem.

Some products are exempt from this electronic lead free requirement. Airplanes and products used in oil refineries are among those exempt.

7.9 Conformal Coatings / Potting

In environments that may expose the electronics to gases, other than air, conformal coating or potting materials may be applied to the PCB surface.

Conformal coating is a material that can be sprayed on the PCB surface. PCBs can also be dipped into a conformal coating liquid. Conformal coatings often cover only the top of components.

Newer atomic layer depository (ALD) coatings show great promise to guard against moisture and oil, but are more expensive then liquid coatings. These may require using sealed packages due to their high penetration ability during application.

Potting is comprised of a material that often flows like water. When it is cured, it becomes a solid similar to a rubber.

Conformal coating can be used to reduce tin whisker problems.

One limitation of using conformal coating is avoiding the use of components that are mounted close to the PCB surface. BGAs can have this problem. Over temperature changes, the conformal coating could expand and cause the IC to pop loose from the PCB pad.

7.10 Liquid Glass

One potential option to deflect tin whiskers is to use something like "liquid glass". Liquid glass would be comprised of a binder and microspheres.

Microspheres are small spheres of glass or ceramic. The diameter could be as small as 20 microns. The glass or ceramic microspheres, when spaced close to each other, could block tin whiskers.

Microspheres are used in car paint. That is why the paint on cars is durable. Car paint might work as a conformal coating, although that remains to be proven.

Chapter 7 Quiz

7-1   Can datasheets be ignored if bench-top testing proves a design to work properly?

7-2   What is the advantage of SPICE models?

7-3   What is the risk of counterfeit components?

7-4   What action can be taken to prevent counterfeit components from being used in an electronic assembly?

7-5   What is a primary advantage of a multi-layer PCB?

7-6   What is a via?

7-7   What is a ground loop?

7-8   What helps eliminate a ground loop?

7-9   Why is a rigid PCB important?

7-10   Why should PCBs be clean?

7-11   Why must solder melt at a lower temperature then most metals?

7-12   What is flux?

7-13   What is a tin whisker?

7-14   What percentage of lead would eliminate tin whiskers?

7-15   What is the primary purpose of conformal coatings and potting?

7-16   What type of component packaging can be negatively affected by a conformal coating?

Chapter 8

ELECTRONICS DOCUMENTATION

8.1 Technical Specifications

Technical specifications document the requirements of an electronic design. Those requirements are based on marketing specifications.

Marketing specifications define what is required to meet the customer needs. Most often, customers know what they want. Sometimes there are new products that customers may not know would be beneficial to them.

It is common for technical specifications to include mechanical size, performance, I/O, temperature range, humidity range, vibration, reliability goal, environment protection, labeling, safety, EMC, and a cost goal.

8.2 Schematics

Schematics depict the configuration of an electronic design. It documents the location of components relative to each other.

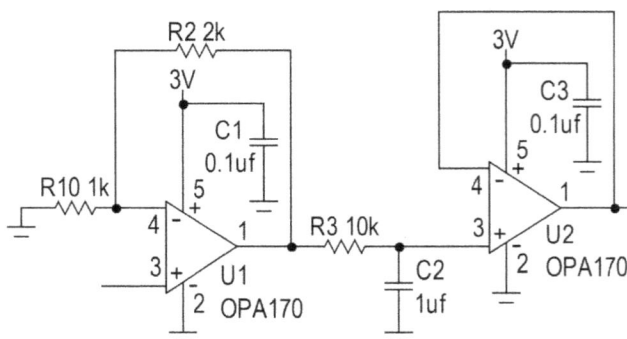

Figure 8.2-1

In Figure 8.2-1, each component has a reference designator (R1, R2, R3, C1, C2, C3, U1, U2). The op amps identify the pin number on the op amp to ensure each connection is correct.

Decoupling capacitors connect the 3V power rail to ground. This is to reduce noise that may be on the 3V power rail.

Often, components span multiple pages. It works well to use TitleCase names as well as the page number where that signal connection can be found (Figure 8.2-2).

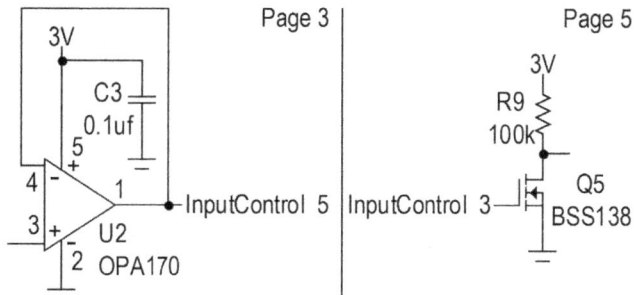

Figure 8.2-2

8.3 BOM (Bill of Materials)

The BOM defines the components for the electronic assembly. It specifies the manufacturer / part number for each component. The part number defines the component and the size.

It is common to list two manufacturers / part numbers on a BOM. This provides a second source if one of the components is not available. Second source components should have no negative effect on product performance. They also help keep component costs lower.

C1	Cap, ceramic, 0.1µF, 10%, 50v, x7r	0603	
	Kemet C0603C104K5RAC_	AVX 06035C104KAT_	
C8	Cap, ceramic, 0.1µF, 10%, 50V, x7r	0603	
	Kemet C0603C104K5RAC_	AVX 06035C104KAT_	
C2	Cap, ceramic, 1µF, 10%, 16V, x7r	0805	
	Kemet C0805C105K4RAC_	AVX 0805YC105KAT_	
C7	Cap, electrolytic, 470µF, 10%, 63V	PTH	
	Nichicon UHE1J471MHD6		
D6	Diode, TVS, 30V, 600w	SMB	
	Microsemi MSMBJ30CA_	Fairchild SMBJ30CA_	
D1	Diode, TVS, 36V, 1500w	SMC	
	ST Microelectronics SM15T36CA	Vishay 1.5SMC36CAHE3_	
D7	Diode, Schottky, BAS70 triple	SOT363	
	Diodes Inc. BAS70TW-7-F	NXP BAS70VV	
D2	Diode, Schottky, BAS70	SOT24-3	
	Infineon BAS70TW	OnSemiconductorBAS70LT1G	
D5	Diode, Zener, 6.8V, 2%	SOT24-3	
	On Semiconductor BZX84B6V8LT1G	Vishay BZX84B6V8-GS08	
C1	Connector, 2x5, smt	FTSH-105-01-f-dv-k	
	Samtec FTSH-105-01-F-DV-K		
L2	Inductor, smt, 33uH	CC_LPS6225	
	Coilcraft LPS6225-333MRB		

L8	Inductor Common Mode, smt, 2mH	50225C
	Murata 50225C	KOA SLF0905TTEB202Y
Q6	FET, BSS84 dual, P Channel	SOT65
	Diodes Inc BSS84DW-7-F	NXP BSS84AKS
Q9	FET, BSS138 dual, N Channel	SOT65
	Diodes Inc BSS138DW-7F7	
Q5	FET, BSS138, N Channel	SOT24-3
	On Semiconductor BSS138LT1G	Fairchild BSS138
Q1	FET, BSS159N, N Channel Depletion	SOT24-3
	Infineon BSS159N	
Q3	FET, IRFU1018, N Channel	TO-251AA
	International Rectifier IRLU1018	
R8	Res, smt, 1k, 1%, 1/10w, 100ppm	0603
	Bourns CR0604-FX-1001_LF	KOA RK73H1JT_1001F
R2	Res, smt, 2k, 0.1%, 1/10w, 100ppm	0603
	Vishay CRCW06032K00FKE_	Panasonic ERA3AEB2001V
R3	Res, smt, 10k, 1%, 1/10w, 100ppm	0603R
	Bourns CR0604-FX-1002_LF	KOA RK73H1JT_1002F
R9	Res, smt, 100k, 1%, 1/10w, 100ppm	0603R
	Vishay CRCW0603100KFKE_	Panasonic ERA3AEB1003V
U5	IC, Voltage Supervisor, 2.9V	SC70-3
	Intersil ISL88002IE29Z-T	Analog Devices ADM803SAKS
U4	IC, Voltage Regulator, ±7V	SOT24-5
	Linear Technology LT1615IS5-1#	
U3	IC, Voltage Regulator, 3.3V	MSOP16
	Linear Technology LTC3630AIMSE#	
U8	IC, RS-485 Transceiver, MAX3483E	SO8
	Maxim MAX3483ESA	Intersil ISL83072EIBZA
U15	IC, FRAM, 512 x 8	SO8
	Ramtron FM25CL64B-GTR	
U11	IC, FLASH, 2M x 8	SO8
	Micron / Numonyx M25P16	Winbond W25Q16
U6	IC, Analog Switch	SOT65
	Texas Instruments 74LVC1G3157DCK	Maxim MAX4730EXT_
U9	IC, Current Monitor, Gain of 50	SC70-6
	Texas Instruments INA213DCKR	
U7	IC, Micro, MSP430F5529	QFN50P400X
	Texas Instruments MSP430F5529IPNR	
U1	IC, Op Amp	SOT24-5
	Texas Instruments OPA170AIDBVR	
U2	IC, Instrumentation Amp	MSOP8
	Texas Instruments INA826AIDGKR	

Some companies create their own part number for each component. For example, the part number 48204523 could represent a 1% 1k resistor with a package size of 0603 and 100ppm. This part number could include a Vishay CRCW06031K00FKE_, KOA

RK73H1JT_1001F, Panasonic ERA3AEB1001V, and Bourns CR0604-FX-1001_LF.

Maintaining internal part numbers for every component adds considerable time and cost. It is much easier to use the component manufacturers' part numbers.

8.4 PCB Layouts

PCB layouts define the path of the metal traces on a PCB. These metal traces connect the components.

It is common to generate a netlist from a schematic. This defines the connections between components. For example, R21 is connected to C4 and to U1 pin 3.

A PCB layout starts with a mechanical requirements drawing. A mechanical requirements drawing describes the size of the PCB and may include the location of connectors, specific components, and holes for mounting screws.

The next step is to load a schematic netlist into a PCB layout program. This defines the components and the connections.

After the PCB layout is complete, the end result PCB netlist is identical to the schematic netlist. This leads to the creation of PCB fabrication documents. Gerber is a common format. These docs include the definition of each component location, each trace, each via, and each test pad, for every PCB layer.

A solder mask places a coating on the PCB which covers the entire PCB, except for places that will be soldered.

A silkscreen is printing on the top and bottom of the PCB to identify the location of components. This is an optional step that may be used for prototypes but not in production to reduce costs.

It is common to put the assembly number and revision number with the copper foil used for the traces. This is less expensive than adding a silkscreen.

The next steps are to analyze the PCB for DFM (design for manufacturability) and DFT (design for testability). DFM verifies that the components are spaced appropriately and not too close to PCB edges.

The DFT verifies that all components can be accessed via a probe to confirm that the correct components have been mounted on the PCB.

The final step is to compile all of the relevant docs (schematic, BOM, mechanical requirements drawing, and Gerber files) necessary to build the electronic assembly.

Chapter 8 Quiz

8-1   What is the purpose of a technical spec?
8-2   What is the purpose of a marketing spec?
8-3   What is the purpose of a schematic?
8-4   How is each electronic component labeled?
8-5   What is the purpose of a BOM?
8-6   What does second source mean?
8-7   What is a netlist?
8-8   What is a solder mask?

Chapter 9

ELECTRONICS TESTS

9.1 Design Verification Tests

Design verification tests verify that the electronic design meets the technical specification. They test performance as specified.

9.2 EMC Tests

EMC (electro-magnetic compatibility) tests verify that electronic products meet world standards. The EMC standards and tests were established by the U.S. Government and by other governments throughout the world. Most of them, fortunately, have similar standards.

The purpose of the EMC tests is to verify that the electro-magnetic fields and signals, created by the electronic equipment, do not interfere with other electronic equipment. They also verify that the equipment can tolerate external EMI (electro-magnetic interference).

The sources of EMI include natural sources such as lightning, intentional sources such as radio transmitters and garage door openers, and unintentional sources such as electric power lines.

The requirements vary depending on the product use. The wide range of usage includes domestic, commercial, industry, vehicle, military, and aerospace environments.

A summary of the requirements includes:

- *Radiated Emissions test verifies that electro-magnetic frequencies generated by the DUT (device under test) do not affect other electronic equipment. The EMI could be generated by items such as clocks, switching power supplies, fast changing pulses, and motors. Common commercial test frequencies are from 30MHz to 1GHz. This test involves the use of a spectrum analyzer, or equivalent, connected to a wideband receiving antenna.*

- *Radiated Immunity test determines if wireless radio frequencies affect the DUT performance. Common sources include handheld radios, vehicle radios, and RF (radio frequency) heaters. Sweeping the appropriate frequency range with a transmitting antenna provides the source for the EMI. The DUT is monitored during the test.*

- *Conducted Emissions test verifies that energy, propagated through power cables and / or data cables by the DUT does not affect other electronic equipment. This test involves the use of a spectrum analyzer, or equivalent, connected to the cables.*
- *Conducted Immunity test determines if external energy, propagated on the DUT power cables and / or data cables, affects the DUT performance. Common sources include RF equipment and power transient equipment. The test is comprised of sweeping the appropriate frequency range by direct connection to the power cables or by capacitive / inductive coupling to the data cables. Transient voltages could be as high as 4,000V. The DUT is monitored during the test.*
- *ESD (electro-static discharge) immunity determines if electric shocks, applied to the DUT housing, affect the DUT performance. The most common source is electric discharge from humans, which can occur in low humidity environments. Other common sources are conveyer belts and high velocity air blowers. This test is accomplished by an ESD gun which generates high voltages, and then discharges them upon contact with the DUT. The voltages could be as high as 15,000V. The DUT is monitored during the test.*
- *Fast Transient Burst Immunity determines if low energy transient (60MHz) arising from disconnect of an inductive load affects the DUT performance. The voltages could be as high as 4,000V.*
- *Surge Immunity determines if high voltage spikes at 300kHz, on the wire connections to the DUT, affect the DUT performance. The voltages could be as high as 6,000V. One source for this occurring after installation at a customer site is lightning.*
- *Magnetic Field Immunity determines if magnetic field noise affects the DUT performance. This affects very few devices. Those affected include electron microscopes and Hall sensors. Reduction*

of receiving loop areas, such as twisted pair wires, is usually effective.

Common EMC specifications are the IEC61000 and CISPR 55011. CISPR is a subcommittee of IEC and specifies only emissions. Immunity is specified by IEC.

The CISPR documents are ultimately merged into the IEC immunity documents to provide both emissions and immunity requirements. Most countries that adopt IEC requirements inherently include CISPR requirements.

The EMC tests are self-certification tests. Testing with the appropriate equipment is acceptable.

As the appropriate equipment may be expensive, there are EMC test companies who have this equipment and are capable of testing per the EMC specifications.

9.3 Firmware Verification Tests

Firmware verification tests verify that the firmware code performs correctly for all situations, including all input control commands. It also verifies that the firmware code handles input errors.

This is a vigorous test that attempts to test every variation to ensure that nothing fails after released to customers.

Good industry practice is to develop these tests as small blocks that are clearly defined. They are often structured so they can be sequenced to "fully" test the product.

Whenever changes are made to the firmware code, full regression testing is necessary to verify the changes do not have an unintended, negative effect on product performance.

Automation of firmware qualification testing removes the tedium of repeating all of the tests every time changes are made to the firmware code.

Chapter 9 Quiz

9-1   What is the purpose of a design verification test?
9-2   What is varied for a design verification test?
9-3   What is the purpose of EMC tests?
9-4   Are EMC tests self-certified?
9-5   What is the purpose of a firmware verification test?
9-6   Why is a firmware verification test important?

Chapter 10

Electronics Manufacturing

10.1 Electronic Components

A manufacturing company attempts to ensure that the correct electronic components have been received. The first step is to verify that the components are correct by checking the component container label. Sometimes items, such as mechanical parts, may require an audit process to ensure that the part meets the specifications.

A second step is to verify the supply chain to eliminate counterfeit components. The supply chain includes information about every company that handled the components starting with the component manufacturing company.

The IPC (International Printed Circuit group) has standards that define component packages, PCB fabrication (IPC600), workmanship (IPC610A), ESD models, flux types, and other important specifications for high quality production of electronic assemblies. They are crucial considerations during the electronic design process. See www.ipc.org for details.

10.2 Kitting

Kitting gathers components together before being placed into the PCB build process. This ensures that all components have arrived.

This also makes it easier to schedule PCB assemblies. Kitting usually occurs a week prior to building the PCB assemblies.

10.3 ESD Boxes / Gloves

As ESD (electro-static discharge) can damage components, it is important to handle the assemblies with gloves, and to transport the assemblies between manufacturing stations in ESD boxes.

The gloves also protect against contamination after the assemblies have been cleaned.

It is important to ensure that the manufacturing process certified ESD standards (both machine and human body) meet the product's ESD requirements.

10.4 SMT (Surface Mount Technology)

A significant development in the miniaturization electronic components is SMT. It has been widely used for electronic assemblies since the late 1980s.

The advantages of SMT components are that they are smaller, and they do not interfere with SMT components mounted on the other side of the PCB. PTH components are larger and consume real estate on both sides of the PCB.

10.5 SMT Equipment

The standard procedure for mounting SMT components to a PCB is to place a solder mask on the PCB. The solder mask is a stencil that has holes for the SMT component pads. A solder paste is wiped across the solder mask so that the PCB component pads end up with solder paste on them.

Newer, direct solder paste print machines are used when it is cost effective.

SMT robotics equipment places SMT components on the PCB. As precision is important, there are fiducials on the PCB that are accurate in relationship to the PCB artwork. These fiducials are small dots of copper foil on the corners of the PCB. The SMT robotics equipment uses the fiducials to ensure the accurate placement of the components on the PCB component pads.

SMT components are on reels or tubes that are mounted to the SMT robotics equipment. SMT robotics equipment is programmed to select components from the correct reel / tube, and place them at the correct location in the correct orientation.

The reels / tubes almost always have labels which identify the components, the component manufacturing companies, and the lot numbers. The lot numbers identify the dates the components were built. All of this information is usually logged so that they can be referenced in the future if the need arises.

After all of the components have been placed, the PCB assembly is placed on a conveyor that takes it through a reflow oven. The reflow oven melts the solder paste which results in the components being soldered to the PCB.

When there are components on both sides of the PCB, the same procedure is followed. The components on the bottom side do not fall off when re-run through the reflow oven. The surface tension of the remelted solder holds them in place.

One common problem can be the solder paste. Solder paste can become dry over time. When it is dry, it may plug some holes in the solder mask. Consequently, some PCB pads end up with no solder paste.

Keeping solder paste in a freezer until used, can reduce the drying problem. Newer machines attempt to solve this problem to control a timeout before the solder paste is used.

The solder mask has holes, most of them rectangular holes. The sides of the holes may be perfectly horizontal, or may be trapezoidal, depending on the size of the hole.

Solder mask and solder paste are crucial to the manufacturing process. The solder mask defines the area which requires solder. Solder paste contains small balls of a solder alloy and flux.

The crucial parameters for a solder mask are the area fill and volume. Both relate to the solder mask stencil ability to cover the solder pads and component leads.

Solder flux must be compatible with the intended cleaning process. IPC guidelines provide many important details.

10.6 Wave Solder

A wave solder machine has a section of molten solder and a conveyer. A PCB, once populated with the relevant components, is placed on the conveyer. The conveyor passes the PCB over the molten solder, makes contact with the molten solder, then continues until the solder is cooled and in a solid state.

This process is often used for soldering PTH components. SMT components on the bottom of the PCB can work fine, if the components are glued to the PCB during the SMT robotics equipment process.

As with SMT reflow, the heat and cool profile is critical to ensure good solder flow without damaging the electronic components.

10.7 Solder Fountain

Sometimes it is not feasible to pass an entire PCB assembly through a wave solder machine. One option is to use a solder fountain.

A solder fountain is a tub of molten solder. The shape of the tub is dependent on the area of the PCB that needs to be soldered for PTH components.

Often a solder mask is used to protect some components. This solder mask is usually a thin layer of polymer.

This can be important, as closely spaced component pads can result in a solder bridge. A solder bridge inadvertently connects two adjacent component pads that should not be connected.

10.8 Hand Solder

When the PCB layout makes it difficult to use wave solder or a solder fountain, the components are hand soldered. Hand soldering is time consuming and labor intensive, which adds to the manufacturing cost.

10.9 Cleaning

It is important to clean the PCB assembly after the soldering process. It is common to spray water on the assembly to remove contaminants. This is a crucial step prior to coating or potting the assembly. Contaminants can result in the inadvertent connection of metal pads or metal traces when exposed to moisture.

One component that is often overlooked is a connector. Often connectors are PTH and are hand soldered. The bottom of a connector often makes solid contact with the surface of the PCB. This can make it difficult to clean under the connector.

ROSE (reduction of solvent extract) and modified ROSE are the most common methods to test electronic assembly production for cleanliness. Ion chromatography is also used.

Depending on the intended life of the product, it may be necessary to set levels lower than industry standards of 10µg/in equivalent chlorine. In terms of reliability, cleaner is better.

10.10 ICT / Flying Probe Tests

An ICT (in-circuit test) is comprised of a hunk of metal or plastic (a fixture) with pogo-pins in the fixture holes. The pogo-pins are spring loaded.

A PCB assembly is pushed down on the fixture with the pogo-pins. The location of the pogo-pins matches test pads on the PCB. Usually all component leads are connected to test pads.

The ICT attempts to check the value of every component for correct value and orientation. It also provides power to the PCB assembly and checks voltages at relevant points. After these steps are complete, the ICT can program micros and FLASH.

To reduce the costs for prototypes, sometimes a flying probe is used to check the value of every component. Often the flying probe cannot check voltages or program micros and FLASH.

10.11 Functional Tests

The number one problem in electronic manufacturing is solder joints. The solder connection may look ok, but a cold solder joint is when it looks ok but is not electrically connected.

A functional test verifies that the assembly was assembled properly. It does not test every option. It does verify that every electronic path is correct and verifies that every signal is correct.

Suppose an assembly communicates with an external computer. Verifying that the computer can communicate with the assembly ensures that section of the assembly is functioning as designed. It is not necessary to verify every command from the computer.

10.12 Conformal Coating / Potting

Conformal coating / potting can make repair of components difficult. Consequently, coating / potting usually occurs after the ICT / flying probe and functional test.

10.13 Post-Pot Tests

Conformal coating or potting could affect assemblies. This process could inadvertently place insulated coatings on electronic contacts of switches and connectors on the assembly. Post-pot testing may be necessary.

10.14 ESD Bags

Assemblies are placed in ESD bags to reduce ESD damage.

10.15 Electronic Contract Manufacturers

Good electronic designed products can go down in flames when electronic manufacturing is assumed to be trivial.

Some companies do not have internal manufacturing capability. Those companies use an ECM (electronic contract manufacturer).

Most ECMs operate on a low profit margin with staff who cannot dedicate themselves to a specific product. If this specific product design is the only one requiring industrial standards, it may be difficult to ensure on-going compliance.

Choosing an ECM may be a corporate decision. However, the electronic design engineer has the ultimate responsibility to ensure that the ECM understands and executes the manufacturing properly (per the required industrial standards).

As the ECM needs to be profitable, the ECM may not employ a procedure specifically tailored to a specific product need.

It is important that the documentation defines the requirements clearly and concisely. IPC standards (Association Connecting Electronics Industries) provide these types of details. It is also important to periodically audit the ECM manufacturing process to verify compliance.

Chapter 10 Quiz

10-1 Why is it important to verify that the electronic components were shipped from the correct location?
10-2 Is an audit process required at the receiving dock?
10-3 What is kitting?
10-4 Why is kitting important?
10-5 Why are ESD boxes important?
10-6 Why are gloves important?
10-7 What are the advantages of SMT components?
10-8 What are fiducials?
10-9 What is the function of a reflow oven?
10-10 What is the purpose of a wave solder machine?
10-11 What is important to ensure good solder flow without damaging the components?
10-12 When is a solder fountain used instead of a wave solder?
10-13 Closely spaced component pads can result in what problem?
10-14 What is the primary disadvantage of hand soldering?
10-15 Why is cleaning important for assemblies which are potted?
10-16 Why can connectors cause a cleaning problem?
10-17 What is a pogo-pin?
10-18 What are the advantages of an ICT vs. a flying probe?
10-19 What is the number one problem in electronic manufacturing?
10-20 Does a functional test attempt to test every electronic connection?
10-21 Why should functional testing occur before conformal coating or potting?

Chapter 11

ELECTRONICS SYNOPSIS

11.1 People Skills

While technical skills are important, people skills are equally important. Learning how to get along with others, not being inadvertently offensive, and being supportive in a team environment are crucial.

Learning how to present to a group is also important. Organizations such as Toastmasters can help develop public speaking skills and help develop good people skills.

Being active in other organizations, beyond engineering organizations, also helps develop people skills. Listing those activities on a resume is good. Some people who interview college students look for this on a resume.

11.2 Learn How to Learn

Of all that is learned in college, "learning how to learn" is the most important. In the world of electronics, technology changes continuously.

When I graduated from college, there were no micros. Computer programming was Fortran on keypunch cards. We used slide rules, as there were no calculators.

I did not realize how much would change. No one mentioned that I would spend a good portion of my time attempting to stay current with new products. Newer products continue to have higher performance, lower power, smaller size, and lower cost.

Staying current with the newest technology is necessary to remain gainfully employed and helping a company remain competitive.

I never took any courses for micros. On my own, I learned the technology, and expanded my code writing from Fortran → Assembler → C → C++ → C#. I knew nothing about wireless or long life batteries, so I learned the latest in these technologies using the Internet.

Knowing how to learn is crucial for electronic design.

11.3 Writing Skills

Good writing skills are important. It is good to strive to have a maximum of three sentences per paragraph. Two sentences per paragraph are better.

This is difficult. It requires considerable effort to use as few words as possible.

The key is to be clear and concise. This increases the odds the document will be read … and understood.

11.4 Code Writing Skills

It is good for electrical engineers to learn code writing skills for micros and for computers. Some companies prefer to hire people who know how to do both; electronic hardware design and write code.

The most common code for micros is C. At times Assembler is useful. Some micro compilers can handle C++.

People who can do both electronic hardware design and write code give the company more flexibility. Depending on the need, people who can do both are more valuable to the company.

11.5 Summary

This book attempts to cover the broad spectrum of electronics in a readable format. Each section could be expanded to provide more details. Additional sections could be included to cover more topics. That is beyond the scope of this book.

The Internet is a good resource that provides detailed information on the topics presented. It is a good second source to extract more information.

There is nothing difficult to learn. Electronics is not complex. Hopefully this book provides some of that insight.

APPENDICES

Appendix A Micro-Cap

Micro-Cap, a high performance SPICE simulator, has many capabilities including identifying outcome changes based on tolerances, temperatures, and worst cases.

Micro-Cap also provides filter designs that are beyond the simple filters mentioned in previous sections. Active filters commonly provide the best filter choices. For example, Micro-Cap provides filter designs for a Butterworth filter, Chebyshev filter, Elliptic filter, and Inverse Chebyshev filter.

Common filters include:

- *Low-Pass filter: blocks frequencies above a specific frequency*
- *High-Pass filter: blocks frequencies below a specific frequency*
- *Bandpass filter: blocks all frequencies outside of a specific frequency range*
- *Notch filter: blocks all frequencies inside of a specific frequency range*

Micro-Cap allows the specification of passband gain, passband ripple, stopband attenuation, passband frequency, and stopband frequency. The end result is a filter design that identifies the necessary components.

Micro-Cap 11 is an integrated schematic editor and mixed analog / digital simulator that provides an interactive sketch and simulate environment for electronics engineers. It has seen eleven generations of refinement since its original release in 1982. Some of the Micro-Cap features include:

- *Multi-page hierarchical schematic editor*
- *Threading for multiple CPUs and faster simulations*
- *Native digital simulation engine*
- *Periodic Steady State analysis*
- *Integral circuit optimizer with multiple optimization methods*

- *Worst Case analysis with RSS, Monte Carlo, and Extreme Value Analysis*
- *Harmonic and inter-modulation distortion analysis*
- *Stability analysis for linear systems*
- *Integrated active and passive filter design function*
- *32000 parts library*
- *Analog and digital behavioral modeling*
- *Schematic waveform probing*
- *On-schematic voltage/state, current, power, and condition display*
- *Dynamic waveforms change while editing*
- *Parameter stepping*
- *Monte Carlo statistical analysis*
- *Measure and performance functions and plots*
- *Animated LEDs, switches, bars, meters, relays, stoplights, and DC motors*

Appendix B LTspice

LTspice IV is a high performance SPICE simulator, schematic capture, and waveform viewer with enhancements and models for easing the simulation of switching regulators.

The enhancements to SPICE have made simulating switching regulators extremely fast compared to normal SPICE simulators, allowing the user to view waveforms for most switching regulators in just a few minutes.

Included with LTspice IV, are macro models for 80% of Linear Technology's switching regulators, more than 200 op amps, as well as resistors, transistors, and MOSFETs.

Appendix C Quiz Answers

Chapter 1 Quiz Answers

1-1   1.5A
1-2   2Ω
1-3   1.5V
1-4   No. Function generators generate AC voltages.
1-5   Hz (Hertz) and cps (cycles per second)
1-6   Function generator

Chapter 2 Quiz Answers

2-1   0603
2-2   ≈10mA
2-3   2k
2-4   2.5k
2-5   5k
2-6   440μs
2-7   ≈7.96Hz
2-8   ≈2258Hz
2-9   Smaller physical size, larger voltage, and non-polarized
2-10   21V
2-11   6.8μF
2-12   33nF
2-13   139.2mH
2-14   A ferrite material
2-15   ≈72.3kHz
2-16   ≈42.5kHz
2-17   Crystal, resonator, and oscillator
2-18   Temperature
2-19   Heat applied when soldered to the PCB can cause damage to the fuse.
2-20   Resettable fuses
2-21   Use a transformer to increase the voltage at the transmission line start, and use a transformer to decrease the voltage at the transmission line end.

2-22   110V AC and 220V AC
2-23   They detect stationary magnetic fields.
2-24   Motors

Chapter 3 Quiz Answers

3-1   1.4k
3-2   The input voltage is connected to the cathode.
3-3   ≈9.1V
3-4   The voltage across the Zener diode increases.
3-5   The current flows into the gate.
3-6   2.9V
3-7   A BJT is controlled by current. An FET is controlled by voltage.
3-8   Gain times the base current equals the collector current.
3-9   LED and phototransistor

Chapter 4 Quiz Answers

4-1   The purpose of a comparator is to compare voltages.
4-2   Hysteresis
4-3   ≈4k
4-4   ≈0.744
4-5   Offset voltage can add error to the output voltage.
4-6   The amount of current into the "+" input and into the "-" input
4-7   The frequency of input voltage which the output voltage can follow
4-8   The speed at which the output voltage can change
4-9   Higher output accuracy and reduced noise
4-10   Constant current is better. Resistance changes in one resistive type sensor could affect the output of another resistive type sensor.
4-11   Common mode rejection ratio
4-12   Exceeding the "+" input relative to the "+" power rail by more than 0.5V can damage the instrumentation amp.
4-13   Linear voltage regulator
4-14   To create an output voltage that is higher than the input voltage, to create a negative output voltage, to create a positive and negative output voltage simultaneously, and power efficiency

4-15   ≈6.87mA
4-16   Noisy power rails
4-17   No
4-18   Monitor the power supply voltage to ensure that it does not drop below a specified voltage
4-19   0V

Chapter 5 Quiz Answers

5-1   To reduce noise problems
5-2   To change the input voltage into a different output voltage
5-3   0x7A
5-4   168
5-5   Allows a number to be twice as large
5-6   Cannot represent negative numbers
5-7   Higher precision
5-8   Doubles the precision
5-9   Converts analog voltages into binary numbers
5-10   ≈2.988V
5-11   Converts binary numbers to analog voltages
5-12   ≈0.15V
5-13   FLASH does not lose its contents when power is removed.
5-14   RAM is faster and can rewrite individual bytes.
5-15   FRAM does not lose its contents when power is removed.
5-16   FRAM requires more real estate and is more expensive.
5-17   A compiler converts human readable code into ones and zeros.
5-18   A PC (program counter) indicates to a micro the next instruction to execute.
5-19   Faster
5-20   Requires more real estate for the additional connections
5-21   A CS enables individual slave components.
5-22   Three plus one CS for each slave
5-23   Two
5-24   I2C communication includes a slave address in SpiMoSi
5-25   There is no master clock.
5-26   Verifies that no data bits have been changed during communication (all data bits are correct)
5-27   RS-422 works better in noisy environments.
5-28   RS-485 can handle multiple devices
5-29   0xE

Chapter 6 Quiz Answers

6-1 EMC (electro-magnetic compatibility) is a branch of electronics about unintentional generation, propagation, and reception of electro-magnetic energy.
6-2 Both
6-3 To limit voltage spikes
6-4 Zener diode
6-5 Current in must equal current out.
6-6 Helps block electronic noise
6-7 Slows rise times
6-8 Inductor

Chapter 7 Quiz Answers

7-1 No. Datasheets provide details about worst case parameters that could occur, regardless of bench-top test results.
7-2 SPICE models can estimate worst case results by varying all component tolerances.
7-3 A counterfeit component often does not perform correctly, especially over temperature variations.
7-4 Purchase from factory authorized sources
7-5 Multi-layer PCBs allow a voltage plane on one PCB layer and a ground plane on a second PCB layer.
7-6 A via is the metal tube connecting different PCB layers.
7-7 Ground loops are multiple paths for a ground connection.
7-8 Limit all connections to one point
7-9 If too flexible, the electronic components and solder connections can crack.
7-10 If not clean, there can be dendrite growth between the metal traces.
7-11 If solder is melted at a high temperature, the metal on components would melt before the solder melts.
7-12 Flux is a chemical cleaning agent that usually removes contaminants to ensure a good solder connection between the component and the PCB pads.
7-13 Tin whiskers are small metal hairs that grow from tin surfaces over time.
7-14 3% lead
7-15 Mitigate exposure to an environment that could damage components
7-16 BGA

Chapter 8 Quiz Answers

8-1 A technical specification documents the technical requirements of an electronic design.

8-2 A marketing specification defines what is required to meet
the customer needs.

8-3 A schematic depicts the components and connections of
an electronic design.

8-4 With a reference designator

8-5 A BOM defines the components for the electronic
assembly.

8-6 Second sources are alternatives that have similar (close
to identical) parameters. They should have no negative
effect on the product performance.

8-7 A netlist defines the connections between components.

8-8 A solder mask is a coating that covers the entire PCB
except for locations which will be soldered.

Chapter 9 Quiz Answers

9-1 A design verification test verifies that the electronic design
meets the technical specification.

9-2 Temperature, humidity, vibration, and extreme I/O levels
which are in the technical specification

9-3 EMC tests verify that electro-magnetic fields created by
the equipment do not interfere with other equipment. They
also verify that the equipment can tolerate EMI.

9-4 Yes

9-5 A firmware verification test verifies that the firmware code
performs correctly for all situations, including all required
inputs and input errors.

9-6 This attempts to test every variation to ensure that nothing
fails after the product is released to customer.

Chapter 10 Quiz Answers

10-1 To eliminate counterfeit components

10-2 Sometimes, when the components are mechanical

10-3 Kitting involves gathering all components before placing
them into the manufacturing process.

10-4 Ensures that all components have arrived

10-5 They decrease the odds of ESD damaging the components.

10-6 Gloves protect against contamination and ESD.

10-7 They are smaller and can be mounted on a PCB with no physical interference with components mounted on the other side.

10-8 Fiducials are used by a SMT robotics machine to ensure the accurate placement of the components on the component pads.

10-9 Reflow ovens melt solder paste which results in the components being soldered to the PCB.

10-10 Mount PTH components quickly

10-11 The heat and cool profile

10-12 When it is not feasible, such as SMT components not glued to the surface, or PTH components need to be soldered on the top and bottom sides of the PCB

10-13 Solder bridges

10-14 Higher cost

10-15 Moisture penetrates potting. Contaminants on a PCB surface, coupled with moisture, can lead to dendrite growth.

10-16 Connector bottoms often make solid contact with the surface of a PCB. This makes cleaning under a connector difficult.

10-17 A pogo-pin is a spring loaded pin. It conforms to the PCB surface and components when the PCB is pushed onto a group of pogopins.

10-18 An ICT can test component values faster, measure voltages, and program FLASH or micro components.

10-19 Bad solder joints

10-20 Yes

10-21 The assembly is easier to test and to repair.

Sources

It is difficult to identify sources for a wide range of topics spanning forty years. Which topics came from which sources are unknown.

Initial information comes from EE faculty (1967-1971) at the University of North Dakota in Grand Forks.

Everything else comes from EE co-workers, datasheets, application notes published by companies who sell electronic components, SPICE models which helped gain insight and led to new ideas, and the Internet.

About the Author

Clyde Eisenbeis is an Electronic Design Engineer. He feels fortunate to have had the opportunity to work on interesting projects.

A native of western North Dakota, he graduated from the University of North Dakota in Grand Forks in 1971. He started his career on radar systems for the U.S. Navy at Texas Instruments in Dallas, Texas.

He worked at 3M in St. Paul, Minnesota. The projects included digital magnetic recording, anti-shoplifting systems, and cochlear implants.

Fisher / Emerson Process in Marshalltown, Iowa, was his most recent company. The projects included industrial controls of valve positioners (pneumatic control and electric control) and wireless monitors (powered by two D-cell batteries with a life of 5 to10 years).

www.ingramcontent.com/pod-product-compliance
Lightning Source LLC
Chambersburg PA
CBHW062029200326
41519CB00017B/4976